RECENT PROGRESS IN SOME AIRCRAFT TECHNOLOGIES

Edited by **Ramesh K. Agarwal**

Recent Progress in Some Aircraft Technologies

http://dx.doi.org/10.5772/61381
Edited by Ramesh K. Agarwal

Contributors

Alberto Ghiraldo, Ernesto Benini, Sebastien Paris, Andreas Marn, Christian Faustmann, Thorsten Selic, Gabriele Neretti, David Johnson, Ignazio Dimino, Rosario Pecora, Francesco Amoroso, Gianluca Amendola, Antonio Concilio, Jawhar Ghommam, Yassine Jimli, Nurradeen Fathallah, Xin Qi, Dalei Song, Zhong Liu, Jianda Han, Liying Yang, Yuqing He, Di Guoqing, Nasim Ullah

CBS, Edition **2017**

Published by InTech

Janeza Trdine 9, 51000 Rijeka, Croatia

Publishing Process Manager Andrea Koric

Technical Editor SPi Global

Cover Designer InTech Design Team

Additional hard copies can be obtained from orders@intechopen.com

Recent Progress in Some Aircraft Technologies, Edited by Ramesh K. Agarwal
p. cm.
Print ISBN 978-953-51-2481-8
Online ISBN 978-953-51-2482-5

INTECH

open science | open minds

Contents

Contents

Preface

This book is a compilation of research articles and review articles describing the latest advances in some engine technologies, active flow control, and morphing technologies and in topics related to aeroacoustics and aircraft controllers. The book is divided into three sections. Section 1 is titled "Engine Technologies." It has two papers dealing with progress in experimental research in turbine aeroacoustics and lubricants for turbine engines. Section 2 is titled "Actuators and Controllers." It has five chapters dealing with a wide variety of topics. The topics include active flow control by using plasma actuators, flight actuators testing and qualifications, morphing technologies (adaptive ailerons), self-healing control design under actuator fault occurrence on single-rotor unmanned helicopter, and a hierarchical tracking controller for quadrotor. Section 3 is titled "Miscellaneous Topics." It has two chapters; one describes studies on effects induced by aircraft noise, and the other is on numerical study of support interferences on SOAR separation wind tunnel test. Thus, the book covers a wide variety of topics related to several aircraft and engine technologies in nine chapters in a single volume. It is hoped that it can serve as a useful source of reference to both researchers and students interested in learning about certain specific aircraft technologies.

Ramesh K. Agarwal
Washington University,
St. Louis, USA

Section 1

Engine Technologies

Chapter 1

Progress in Experimental Research of Turbine Aeroacoustics

Andreas Marn, Christian Faustmann and Thorsten Selic

Additional information is available at the end of the chapter

http://dx.doi.org/10.5772/62431

Abstract

Modifications on the intermediate turbine duct in order to reduce noise emissions by changing interaction frequencies and/or modes capable to propagate are presented. Also different turbine exit casings are described that are optimised to reduce interaction noise that is propagating through the engine and is one of the major noise sources during landing (operating point approach). The most promising modifications to reduce sound power levels are described. Depending on different modifications at specific operating points, the reduction of sound power level is between 5 dB and 10 dB, which is a significant reduction. However, some of these measures show an increase in aerodynamic losses. Therefore, a compromise has to be found between higher losses during a short duration (e.g. landing) and significant noise reduction. The chapter focuses on experimental results obtained in the test facilities of the Institute for Thermal Turbomachinery and Machine Dynamics at Graz University of Technology.

Keywords: aeroacoustics, low pressure turbine, high pressure turbine, turbine exit casing, intermediate turbine duct

1. Introduction

In modern commercial high bypass ratio aero engines, low specific fuel consumption together with low life cycle costs are mandatory to allow air transportation economically. Furthermore, an environmentally friendly aero engine should have reduced CO_2 emission by even lower fuel burn and generate low noise levels during takeoff (side line and cut back) and landing.

At the beginning of the jet era noise emissions were dominated by jet mixing noise. That has been reduced when introducing modern bypass engines by lowering the jet speed. Noise emitted by fan, turbine, and compressor has then become important with the reduction in jet speed. Now, that noise has to be reduced significantly. Since the publication of the ACARE goals, which are targets defined by the Advisory Council for Aeronautics Research in Europe to reduce the environmental impact of air transportation, the commercial and political pressure to reduce CO_2, NO_x and noise (up to 20 dB reduction of noise level until 2020 compared to technologies of the year 2000) has been increased considerably. A lot of research over the past decades was done reducing fan and compressor noise in order to achieve the noise reduction targets. During the last years, noise emission from the fan was much reduced that suddenly noise from the interaction of the last stage low pressure turbine and the turbine exit guide vane of the exit casing became perceivable. Nowadays also manufacturers of low pressure turbine components have to consider acoustic aspects in their design to be able to reach the ACARE goals in 2020. Therefore a lot of research is currently done in that field of expertise. An additional benefit of that low noise levels is that passengers as well as residents living in the vicinity of airports feel more comfortable. Basically an increasing acceptance to live close to airports and rise of life quality can be achieved if the noise level of aero engines is decreased significantly.

In this chapter measures to reduce noise generated and propagating from modern aero engine turbines are presented. The main issue is to find methods and/or new engine designs that reduce noise without causing considerable losses or a reduction of thrust. However, a lot of novel engine architectures are investigated in several national and international funded projects with the goal to reduce the emission of pollutants, e.g. by lowering the engine weight. This can be achieved by reducing the length of the entire engine by reducing the axial spacing between blade rows or integrating additional functions in one part, e.g. a non-lifting strut in an intermediate turbine duct which also has to turn the flow and provide the next rotor with the correct inflow conditions. This leads to a so-called turning mid turbine frame. However, all modifications on these parts of the engine will influence the noise generation and propagation. Also a considerable change of excitation of blades and vanes can be observed.

2. Experimental setups and test facilities

The Institute for Thermal Turbomachinery and Machine Dynamics at Graz University of Technology operates a 3 MW compressor station in order to supply a couple of test facilities continuously with pressurized air. Among them are the subsonic test turbine facility for aerodynamic, acoustic, and aeroelastic investigations (STTF-AAAI) as well as the two-spool Transonic Test Turbine Facility (TTTF).

2.1. Subsonic test turbine facility for aerodynamic, acoustic, and aeroelastic investigations

In the described subsonic turbine test facility for aerodynamic, acoustic and aeroelastic investigations (STTF-AAAI), the maximum pressure ratio is limited to 2 due to the inlet spiral casing. The mass flow rate is limited to 15 kg/s due to the compressor station characteristic. A temperature at stage inlet of max. 100°C can be realised. This inlet temperature can be adjusted within a wide range by means of water air coolers. The pressurized air enters the facility through a spiral inlet casing where the flow is turned into axial direction. That spiral inlet casing also supports the front bearing of the overhung-type turbine shaft. The shaft is coupled to a water brake counteracting the power of the turbine. The necessary cooling water cycle of the brake is connected to the re-cooling plant of the institute.

For a test rig it is mandatory to provide well-defined and uniform inflow conditions; therefore a de-swirler and a perforated plate are located far upstream of the inlet guide vanes. That mentioned inlet guide vanes upstream of the stage (and downstream of the perforated plate) simulate additional wakes of other upstream low pressure turbine stages. The air leaves the test rig through a cylindrical acoustic measurement section, supporting struts centring the acoustic measurement section, exhaust casing, and the exhaust stack to ambient. For more information a detailed description of the subsonic test turbine facility can be found in Ref. [1].

2.1.1. *Turbine stage and turbine exit casings[SEQA1]*

During the aerodynamic design process it was crucial to achieve relevant model parameters to reproduce the full scale low pressure turbine (LPT) configuration. The diameter of the test rig rotor is about half of that of a commercial aero engine LPT. Therefore the rig is operated at higher rotational speeds to realise an engine relevant loading parameter. A meridional section of the rig is shown in **Figure 1**. The state-of-the-art (reference) and the leaned turbine exit casing (TEC) are shown in the sketch at the top (a). The inverse cutoff as well as the high lift design (H-TEC) can be seen at the bottom (b). It has to be mentioned that the bladings are not drawn to scale. The rig is characterised by a high aspect ratio unshrouded low pressure turbine rotor followed by the TEC. Relevant geometry parameters can be seen in the upper half of **Table 1**.

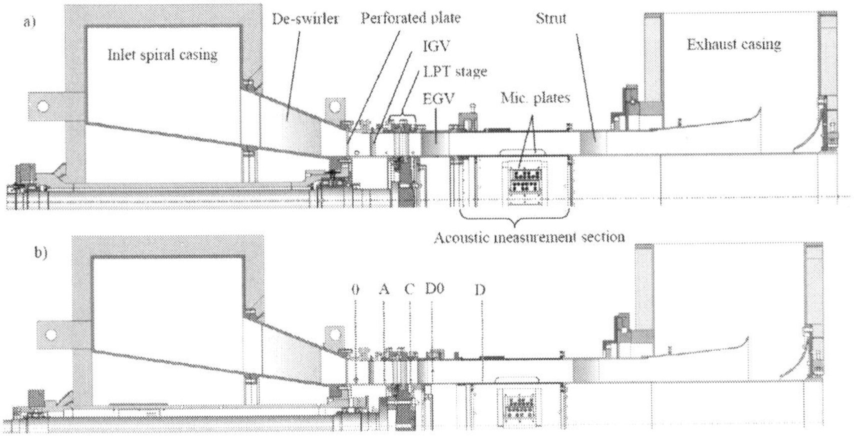

Figure 1. Meridional section of the STTF-AAAI; (a) reference and leaned TEC and (b) inverse cutoff and H-TEC.

Four different TEC setups with different vane counts (see **Table 1**) have been tested but the leading edge is at the same axial position for all configurations. One significant difference is that the reference and leaned TEC are manufactured without fillets while the TECs with smaller chord length have fillets at hub and tip due to manufacturing and assembly requirements. The leaned TEC was optimised (detailed information can be found in Ref. [2]) in order to reduce rotor-TEC interaction noise by keeping the profiles of the turbine exit guide vanes (TEGV) to be able to lead through the same supply lines as through the reference TEC. As it was shown in some European projects, e.g. DREAM, the rear bearing can move forward under the TCF section for future engine architectures giving the designer the opportunity to aerodynamically and/or acoustically optimise the vanes of the turbine exit casing. Therefore, the third setup is an acoustically optimised TEC named inverse cutoff TEC. The basic idea of that setup is to utilise a small cutoff corridor in between two cuton regions. A detailed description can be found in Ref. [3] and a verification and comparison with experimental results is given in Ref. [4]. The fourth setup is aerodynamically optimised and is designed to reduce losses at aero design point.

Further, the rig has some inlet guide vanes (IGV) in order to impose some typical pre-swirl on the flow. Stator vanes are located downstream of the IGVs, followed by the rotor and the turbine exit guide vanes (TEGV). The 1 and 1/2 stage is representative of the last stage of a commercial engine with TEGV. **Table 1** shows the blade count and the main geometrical details of the turbine. The rig in its current setup is characterized by a high aspect ratio unshrouded rotor followed by one of the above described turbine exit casings. The tip leakage flow dominates the flow field downstream of the rotor. The flow through the guide vanes is mainly influenced by secondary flows. The TEGV are designed to turn the swirling flow into an axial direction (reducing swirl and lower the kinetic energy of the flow) and to recover some static pressure.

Additionally, for this test rig, a second stator and low pressure turbine rotor has been designed. Stator and rotor have the same blade count as the reference design, but different profile geometries, including a rotor with a 20% increased loading parameter. The design intent of that second stage was to provide a similar/identical rotor exit flow as well as shaft power of the test rig. Because of the larger turning of that highly loaded rotor, the operating points must have been adjusted. A lower rotational speed in order to keep the power output identical was chosen. However, the stage pressure ratio has been kept the same as for the datum stage. There have been two reasons to keep the blade and vane counts identical. Firstly, it is the geometrical limitation of the test rig. Axial chord of both stator and rotor had to be the same as well as the axial distance between the vanes and blades. Secondly, the resulting acoustic and aerodynamic field is dependent on the number of blades and vanes of the test rig [5]. In order to keep the modal structure of the blade/vane interactions the same as for the datum stage, the number of blades and must be the same.

2.1.2. Operating conditions

Because the STTF-AAAI is used for both acoustic and aerodynamic investigations, the main operating points are selected according to relevant noise certification points. They have been defined using a typical aerodynamic design point of a last stage of an LPT. That design point is derived from current LPT design practice and scaled along reduced speed, reduced mass flow (both referred to 288.15 K and 1013.25 mbar) and pressure ratio. For this investigation the acoustically relevant operating point approach was chosen. The Reynolds number of the TEGVs is defined using the midspan conditions at rotor exit as well as the axial chord of the vanes. The lower half of **Table 1** shows the operating conditions.

Geometry details	
Number of blades/vanes	
IGV/Stator/Rotor/	83/96/72
TEGV state-of-the-art (Reference TEC)	15
TEGV leaned TEC	15
TEGV inverse cutoff (inv.)	>40
TEGV H-TEC	<35
Tip gap to blade height ratio	1.0%
Hub to tip radius ratio	≈2/3
Operating conditions	
TEGV Reynolds number ratio $Re_{ref.}/Re_{H\text{-}TEC.}$	~3
TEGV Reynolds number ratio $Re_{ref.}/Re_{inv.}$	~3
TEGV Reynolds number ratio $Re_{ref.}/Re_{leaned}$	~1
Diffusion factor ref./leaned/inv./H-TEC	~0.5/~0.5/~0.6/~0.6
Reduced mass flow	6.94 kg/s
Reduced rotational speed	3653 rpm/4060 rpm
Stage pressure ratio	1.16
Stage total inlet temperature	100°C

Table 1. Geometry details and operating conditions.

2.2. Transonic test turbine facility

The setups being tested consist of a single-stage unshrouded transonic HP turbine. Downstream of that turbine stage, a S-shaped turbine centre frame, which is the main part of interest for this investigation, is located. The centre frame is followed by a shrouded counter-rotating LPT rotor (see **Figure 2**). The pressurized air flows through the transonic unshrouded HPT rotor and enters the turbine centre frame. The air is then turned by 16 struts in the turbine centre frame (resulting in a turning mid turbine frame [TMTF]) in negative direction relative to the rotation of the HP rotor. After that the air enters the LP rotor at a larger diameter and with an appropriate swirl angle. Blading parameters and operating conditions can be seen in **Table 2**.

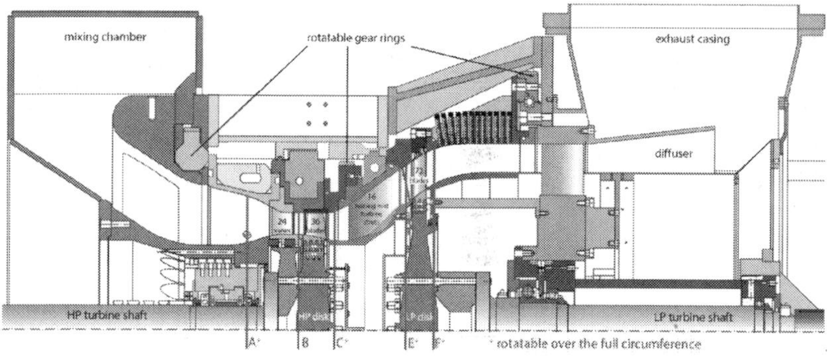

Figure 2. Meridional section of the TTTF.

Blading parameters					
	HP vane	**HP blade**	**Strut C1/C2/C3**	**Splitter C3**	**LP blade**
Vane/blade no.	24	36	16	32	72
h/c_{ax}	1.15	1.37	0.46/0.53	3.5	2.94
Tip gap	- Unshrouded -			-	Shrouded
Operating conditions ADP					
	C1	C2		C3	
BPF_{HP} $[kHz]$	6.69	6.69		6.69	
BPF_{LP} $[kHz]$	4.26	4.26		4.26	
Stage p_t ratio HPT/ LPT	3/1.3	3/1.3		2.83/1.36	
Power [MW] HPT/ LPT	1.44/ 0.3	1.44/ 0.3		1.43/ 0.3	

Table 2. Blading parameters and operating conditions.

Configuration 1 (C1) consists of 16 turning struts. It has a non-dimensional length of about 3.5 (Lax/hin) and an area ratio of 2. C1 was designed using a quite complex three-dimensional design of the strut and keeping rotationally symmetric endwall contours. The struts have a maximum thickness to chord ratio of 22% at about 25% of the axial chord length to provide enough space for service lines like oil pipes and for load carrying structures. Two major goals were set for the aero design of the second configuration (C2). Firstly, C2 has to be more aggressive than the first configuration and secondly, the LPT performance has to be unchanged, keeping the same pressure loss. For C2 also, 16 turning struts have been assembled. For all configurations the same high pressure stage and low pressure rotor have been used.

Therefore, the radial offset and the area ratio are the same for both setups but configuration C2 was designed to be 10% shorter than configuration C1. C2 has a non-dimensional length of 3.1. Close to the hub the axial gap between the strut trailing edge and the leading edge of the LPT is the same as for C1. However, that gap is 20% shorter at midspan and 50% at the casing, respectively. The vane is designed to produce minimum blockage and therefore, it should create minimum losses. Both struts give the possibility to lead through identical service lines. To avoid additional losses and provide same inflow conditions to the LP turbine as the configuration C1, non-axisymmetric endwall contouring was applied at the hub. The optimisation of endwall contour was performed using parameterization based on orthogonal basis perturbation functions. The third setup is based on the same geometry as C1, but 32 splitter vanes have been added for setup C3. Within each strut passage two non-lifting splitters are located. The splitter vanes have been numerically designed and represent a compromise between aerodynamic effectiveness and additional blockage, because it was required to use the same casing parts. Therefore, it was not possible to change the area to count for the additional blockage of the splitters.

2.3. Instrumentation

Five-hole probes with a probe head of 2.5 mm diameter have been applied. The probes were used in measurement plane C downstream of the rotor and in plane D downstream of the TEGV (see **Figure 1b**). The probes are calibrated in a Mach numbers range between 0.1 and 0.8 in 0.1 steps, yaw angles between –20°and +20° in steps of 4 deg, pitch angles between – 16°and +20°, also in steps of 4°. Negative values of the yaw angle indicate a counter-rotating flow and negative values of the pitch angle indicate the flow direction towards the hub. A multi-parameter approximation correlates the calibration characteristics and the flow value to be measured.

The axial positions of measurement planes (marked with letters) can be seen in **Figure 1** for the STTF-AAAI and **Figure 2** for the TTTF. TEGV/Strut inlet plane is located downstream of the rotor trailing edge. TEGV exit plane can be found 55% of the axial chord length of the H-TEC TEGV downstream of its trailing edge and also 130% axial chord length of the reference TEC downstream. The measurement grid covers one TEGV or a strut pitch and about 95% passage height. The five-hole probe was traversed along radial lines. In each measurement point the probe has been aligned with the flow vector to reach the highest accuracy and further to ensure not to exceed the calibration range of the probe (with these probes, it would not have been necessary if one can ensure to be always within the calibration range).

In order to calculate the sound power (propagating downstream) from measured sound pressures, 12 flush mounted condenser microphones (1/4″) at the hub and 12 at the casing have been applied. The microphones have been located in the rotatable acoustic measurement section that can be rotated 360° with arbitrary step size. In addition to these microphones one additional microphone was mounted at a stationary position downstream of the TEGV's as well as struts trailing edge and is used as a reference. The complete sound field was detected at the hub and the casing by traversing the section 360° in steps of 2°. Some more detailed information about the acoustic measurement section is given in Moser et al. [6] and in Faustmann et al. [7].

2.3.1. Measurement uncertainty

The measurement system is made up by 11 multichannel pressure transducers PSI 9016 with a total amount of 176 channels and an accuracy of 0.05% full scale for pressure measurements. Four National Instruments Field Point FP-TC-120 eight-channel thermocouple input modules and one FP-RTD-122 resistance thermometer input module is used. **Table 3** shows the measurement uncertainties (within a 95% confidence interval) of the five-hole probe measurements. These values contain an error due to the multi-parameter approximation, random error and the systematic error of the pressure transducers. The difference between the positive and the negative direction is a result of the multi-parameter approximation. The measurement uncertainties of the static pressure and the total pressure at test rig inlet as well as at stage inlet are ±1 mbar. Uncertainties for total pressure measurements up- and downstream of the TEGVs are also in the range of ±1 mbar. The overall uncertainty of the total pressure loss coefficient ζ is estimated to be about ±0.0014. The random fluctuation of rotational speed is below 0.2% of the current operating speed. Measurement uncertainty of the temperature measurement is about ±0.5 K. The day-to-day variation of the operating parameters such as pressure ratio, corrected speed, rotational speed, total pressure and temperature at rig inlet has been below 0.5%.

Ma	+0.006	-0.003	[/]
α	+0.5	-0.08	[°]
pt	+3.3	-3.0	[mbar]
p	+5.3	-5.2	[mbar]

Table 3. Measurement uncertainties of the five-hole probe.

3. Acoustic analysis theoretical background

The acoustic analysis consists of four constitutive parts:

- phase averaging and adaptive resampling;
- modal decomposition;
- azimuthal and radial mode analysis; and
- computation of sound power.

After performing a Fast Fourier Transformation the sound pressure can be described as sum of acoustical modes propagating along the duct at a specific frequency. These modes are able to propagate in circumferential as well as in radial direction depending on the cuton frequency.

3.1. Phase averaging and adaptive resampling

In order to determine the acoustic effects a phase locked averaging is done. For the two-spool rig the phase of one of the two rotors is chosen. A shaft encoder from the monitoring system generates a pulse per revolution signal indicating start and end of one revolution. Triggering the flow is performed according to the triple decomposition procedure, which characterize a single source of periodic unsteadiness [8].

$$p(t) = \bar{p} + \langle p \rangle + p'(t) \tag{1}$$

The time dependent pressure $p(t)$ is composed as sum of the averaged pressure, the purely periodic component p associated with a coherent periodic structure and the random fluctuation $p'(t)$ that is mainly associated with turbulence.

Each revolution is divided into a fixed number of samples in order to correct small rotational speed variations of the rotor shaft [9]. The average of the samples at the same phase gives the phase averaged values. That procedure is well established and allows the identification of structures correlated to the rotor rotational speed. For a two-spool rig all fluctuations of flow quantities induced by the other rotor are removed. Also, depending on which trigger signal is used for the analysis acoustic effects from the HP-rotor or the LP-rotor can be determined.

3.2. Modal decomposition

The acoustic field at any circumferential position can be written as superposition of several space and time dependent sound pressure waves (acoustic modes). The propagation of that pressure waves is described by the linearised wave equation $\frac{1}{c^2}\frac{\partial^2 p}{\partial t^2} - \Delta p = 0$. The solution of that equation is represented by the general expression (given in many publications, e.g. [10–14]):

$$p(x,r,\varphi,t) = \sum_{m=-\infty}^{\infty} \sum_{n=0}^{\infty} \left(A_{mn}^{+} e^{ik_{mn}^{-}x} + A_{mn}^{-} e^{ik_{mn}^{-}x} \right) f_{mn}\left(\sigma_{mn} \frac{r}{R} \right) e^{im\varphi} e^{i\omega t} \tag{2}$$

A_{mn}^{+}, A_{mn}^{-} are the complex modal amplitudes of order (m,n). k_{mn}^{+} and k_{mn}^{-} represent the axial wave numbers, ω is the angular frequency. The modal shape factor f_{mn} depends on the eigenvalues of the Bessel function σ_{mn} and the geometry given by the hub-to-tip-radius ratio $\frac{r}{R}$. f_{mn} represents the solution of the Bessel differential equation, describing the radial acoustic field considering hard wall boundary conditions [15, 16] and is defined as:

$$f_{mn}\left(\sigma_{mn}\frac{r}{R}\right)=\frac{1}{\sqrt{F_{mn}}}\left(J_{mn}\left(\sigma_{mn}\frac{r}{R}\right)+Q_{mn}Y_{mn}\left(\sigma_{mn}\frac{r}{R}\right)\right) \quad (3)$$

J_{mn} and Y_{mn} are the Bessel functions of mth order of first and second kind. Q_{mn} is also an eigenvalue and F_{mn} is a normalisation factor transforming the system from an orthogonal to an orthonormal eigensystem [15, 16].

The axial wave numbers depend on the local flow properties, such as axial Mach number Ma_{ax} and the swirl number Ω. For the following investigations the swirl is approximated by a rigid body rotation of a steady flow, which leads to a modification of the wave number definition [17]: $\tilde{k}=k-m\frac{\Omega}{c}$. The axial wave numbers are then calculated as follows:

$$k_{mn}^{\pm}=\frac{\tilde{k}}{1-Ma_{ax}^{2}}\left[-Ma_{ax}\pm\sqrt{1-\left(1-Ma_{ax}^{2}\right)\frac{\sigma_{mn}^{2}}{\left(\tilde{k}R\right)^{2}}}\right] \quad (4)$$

Physical and geometrical conditions allow only the propagation of a certain number of specific mode combinations (m,n) along the duct. The axial wave number $k_{mn}^{\pm}$ has to be real; otherwise the result of Eq. (2) will yield to an exponential sound pressure decay if $k_{mn}^{\pm}$ is a complex number. The frequency at which a mode (m,n) is first able to propagate is defined by the cuton frequency:

$$(\mathrm{kR})_{\mathrm{mn}}=\sqrt{1-\mathrm{Ma}_{\mathrm{ax}}^{2}\sigma_{\mathrm{mn}}}+\frac{\mathrm{m}\Omega\mathrm{R}}{\mathrm{c}} \quad (5)$$

The swirl factor $\frac{\mathrm{m}\Omega\mathrm{R}}{\mathrm{c}}$ shifts the cuton frequency to higher or lower values, depending on the sign of Ω. Or in other words, for a specific frequency modes to be cut on are also shifted to higher or lower azimuthal mode orders m, hence the propagating mode distribution becomes asymmetric. The specific modes propagating through a duct downstream of a turbomachine stage result from the rotor-stator interaction and are specified by a simple mathematical relation proposed by Tyler and Sofrin [5]:

$$\mathrm{m}=\mathrm{hB}\pm\mathrm{kV};k=\cdots,-2,-1,0,+1,+2,\cdots \quad (6)$$

h represents the harmonic index (e.g. 1 for the first blade passing frequency, 2 for the second, etc.), and B and V are the number of rotor blades and the number of stator vanes, respectively. According to Eq. (1) it is possible to determine the interactions of the rotor with a complete vane assembly by simply superimposing the effect of the single event. For a stator-rotor-stator assembly the modes can be predicted easily when extending Eq. (6).

$$m = hB \pm k_1 V_1 \pm k_2 V_2; k_{1,2} = \cdots, -2, -1, 0, +1, +2, \cdots \tag{7}$$

3.3. Azimuthal and radial mode analysis

The computation of the propagating sound field in a turbomachine is based on the determination of the amplitudes $A^{\pm}_{mn}$ in Eq. (2). That means that the sound pressure at several axial and circumferential positions has to be measured first, e.g. by means of a microphone array. Then the radial mode analysis (RMA) described, e.g. in Holste and Neise [13], Tapken and Enghardt [15], and Enghardt et al. [12, 18] is applied. The RMA is a methodology used for the modal decomposition of an in-duct acoustic field. The data is used to reconstruct at specific times the instantaneous circumferential pressure distribution. The application of a spatial Fourier Transformation over this data set represents the first action and is called azimuthal mode analysis (AMA):

$$A_m(x, r) = \frac{1}{N_\varphi} \sum_{k=1}^{N_\varphi} p(x, r, \varphi_k) e^{-im\varphi_k} \tag{8}$$

In order to determine the complex amplitudes $A^{\pm}_{mn}$ at a specific angular frequency ω for each azimuthal mode m Eq. (2) can be written as [18, 19]:

$$A_m(x, r) = \sum_{n=0}^{\infty} \left(A^{+}_{mn} e^{ik^{+}_{mn}x} + A^{-}_{mn} e^{ik^{-}_{mn}x} \right) f_{mn}\left(\sigma_{mn} \frac{r}{R} \right) \tag{9}$$

For each azimuthal mode order m a linear equation system $\mathbf{A}_m = \mathbf{W}_m \mathbf{A}_{mn}$ can be found. Since this equation system is (usually) highly overdetermined, a least-mean-square fit algorithm is used to solve that inverse problem.

3.4. Computation of sound power

Considering the energy carried by each individual mode in a hard walled duct, the effective sound power can be determined according to Morfey [17]:

$$A_m(x, r) = \sum_{n=0}^{\infty} \left(A^{+}_{mn} e^{ik^{+}_{mn}x} + A^{-}_{mn} e^{ik^{-}_{mn}x} \right) f_{mn}\left(\sigma_{mn} \frac{r}{R} \right) \tag{10}$$

The complex factor $\alpha_{mn} = \sqrt{1 - \left(1 - Ma^2_{ax}\right) \frac{\sigma^2_{mn}}{(kR)^2}}$ contains the definition of the cuton frequency.

4. Research on TECs

Figures 3 and **4** show the sound power level (PWL) of different azimuthal mode orders m that are cuton. The PWL is evaluated for the first blade passing frequency (BPF) at the operating point approach. The black bars in both figures indicate the sound power level of sound waves propagating in flow direction PWL^{+} and the gray bars show the PWL of propagating waves against flow direction PWL^{-}. The dotted lines in the figures indicate the noise level of the measurements. The microphones are mounted on a vibrating machine which causes some systematic error.

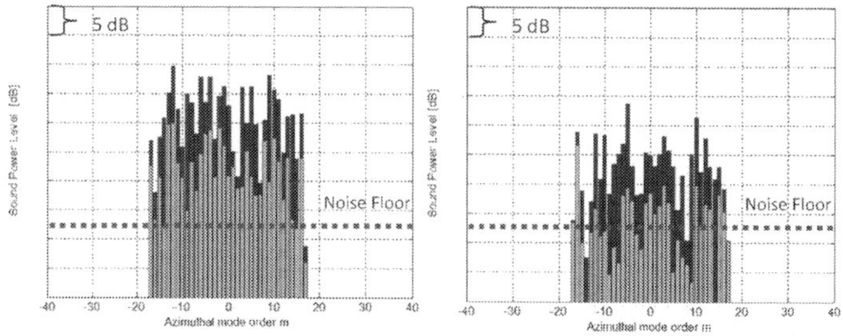

Figure 3. Sound power level of the reference TEC (left) and the leaned TEC (right).

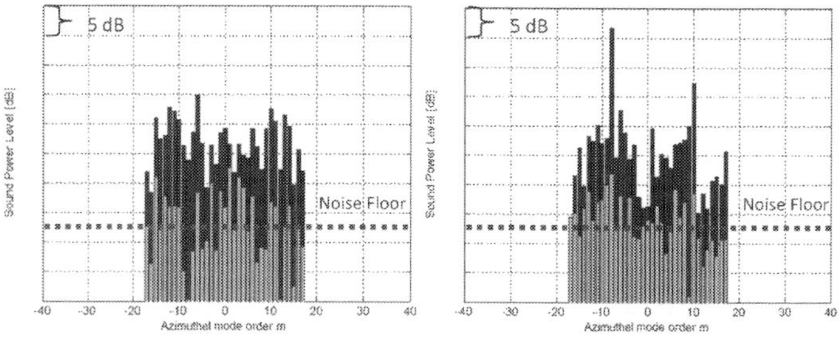

Figure 4. Sound power level of the inverse cutoff TEC (left) and the H-TEC (right).

Together with noise generated due to the evaluation technique and noise related to the measurement error the depicted level of overall noise floor can be assumed. **Figure 3** (left and right) shows the sound power level for the reference TEC and the leaned TEC, respectively. For the leaned TEC it can be seen that the overall PWL is about 8 dB lower than that of the reference TEC. Comparison of the inverse cutoff TEC (**Figure 4** (left)) with the reference TEC shows that $PWL^{\pm}$ is about 5 dB lower for the inverse cutoff TEC. Figure 3 (left) and **Figure 4** (right) show again the sound power level of different azimuthal mode order m that are cuton for the two TEC configurations (reference and H-TEC). Both are again evaluated for the first

BPF at operating point approach. The sound power level of the H-TEC configuration is about 1 dB lower than for the reference TEC. The noise reduction levels mentioned above also contain non-representative modes that belong to interaction modes with certain test rig components. These modes are created for example due to interaction of the sound field and the rear supporting struts at rig exit resulting in additional scattered modes. The difference in overall sound power level of the first BPF $\Delta PWL = PWL - PWL_{ref}$ in and against flow direction as well as the sum of both is also given in **Table 4**.

Configuration	ΔPWL^{+}	ΔPWL^{-}	$\Delta PWL^{\pm}$
Reference TEC	0.0 [dB]	0.0 [dB]	0.0 [dB]
Leaned TEC	-8.1 [dB]	-7.8 [dB]	-8.1 [dB]
Inv. cutoff TEC	-5.2 [dB]	-8.2 [dB]	-5.4 [dB]
H-TEC	-0.6 [dB]	-8.6 [dB]	-1.0 [dB]

Table 4. Overall sound power level change for 1BPF.

Evaluating only engine relevant specific airfoil interactions shows a different trend than that depicted in **Table 4**. **Table 5** shows the azimuthal airfoil interaction mode orders (cuton modes) and also scattered modes calculated according to the equation of Tyler and Sofrin [5]. Mode orders higher than ±17 are cut off. The main interaction mode of the rotor with the TEGV of the reference TEC, leaned TEC, and the H-TEC is cuton and clearly visible (see **Figures 3** and **4**).

Configuration	IGV-Rotor	Vane-Rotor	Rotor-TEC
Reference TEC	-4, +11	-6, +9	-12, +3
Leaned TEC	-4, +11	-6, +9	-12, +3
Inverse cutoff TEC	+11	Cutoff	Cutoff
H-TEC	+11	Cutoff	Cuton

Table 5. Airfoil interaction modes m.

Table 6 shows how the sound power level changes (compared to the reference TEC) when considering only the main airfoil interaction modes given in **Table 5**. It can be seen that the inverse cutoff TEC achieves the highest reduction in sound power level (14 dB) of all investigated configurations. The leaned TEC reduces PWL by about 11 dB, but that means that it is still twice as "loud" as the inverse cutoff TEC. The aerodynamically optimized H-TEC even increases the sound power level by about 2 dB. At aero design point both TECs, the reference TEC and the leaned TEC have the same PWL. The aerodynamically optimized H-TEC is about 1 dB louder than the reference TEC and the inverse cutoff TEC reduces PWL by about 3 dB which is half as loud as the reference TEC.

Configuration	ΔPWL^{+}	ΔPWL^{-}	$\Delta PWL^{\pm}$
Reference TEC	0.0 [dB]	0.0 [dB]	0.0 [dB]
Leaned TEC	-11.3 [dB]	-9.4 [dB]	-11.2 [dB]
Inverse cutoff TEC	-13.8 [dB]	-17.7 [dB]	-14.0 [dB]
H-TEC	+2.4 [dB]	-11.5 [dB]	+2.0 [dB]

Table 6. Sound power level change for 1BPF (airfoil interaction modes).

Considering only modes due to the TEC interaction and the scattering of modes at the TEC it can be seen that the inverse cutoff TEC even reduces PWL by 94 dB(!), because that mode is cut off. The leaned TEC has a reduced PWL of 11 dB and the H-TEC is again increasing the PWL by about 2 dB. In addition to the acoustic measurements five-hole probe measurements have been performed in order to compare the aerodynamics (loss estimation) of the three different TEC configurations with that of the state-of-the-art reference TEC.

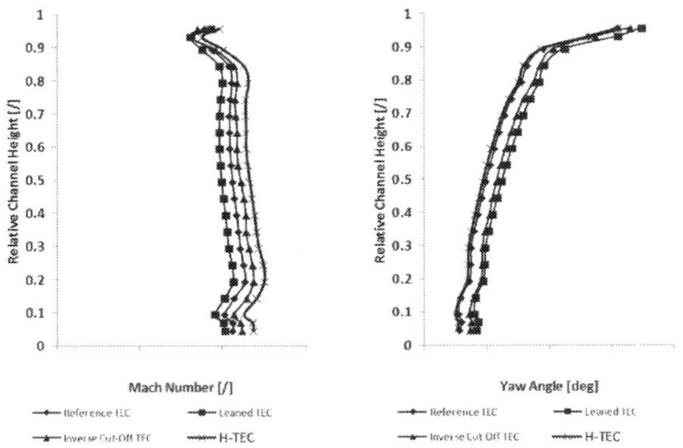

Figure 5. Mach number distribution (left) and yaw angle distribution (right) at TEC inlet.

4.1. TEGV inlet flow: plane C

Comparing the TEGV inlet conditions of the different measurement campaigns for the different TEC configurations reveals only minor differences in flow quantities due to the upstream potential effect of the turbine exit guide vanes of TEC. The circumferentially mass averaged radial Mach number distribution show insignificant differences between the configurations (**Figure 5** (left); 1 tick mark=0.1). Also the yaw angle is almost the same for all three configurations as can be seen in **Figure 5** (right) (1 tick mark=20°), the difference is also negligible. The largest difference between the TEC configurations can be seen at the tip above 90% channel height.

4.2. TEGV outlet flow: plane D0, D

Figure 6. shows Mach number and yaw angle distribution downstream of the TEGVs. H-TEC and inverse cutoff TEC show a much more uniform distribution of the yaw angle at TEGV exit than the leaned and the reference TEC due to the higher vane count, thus, significantly improving the inflow conditions to a following component such as a mixer. Due to the fact that the inverse cutoff and the H-TEC does not consider additional blockage because of the higher TEGV count the Mach number in plane D could be possibly slightly higher than for the reference and the leaned TEC (same TEGV count).

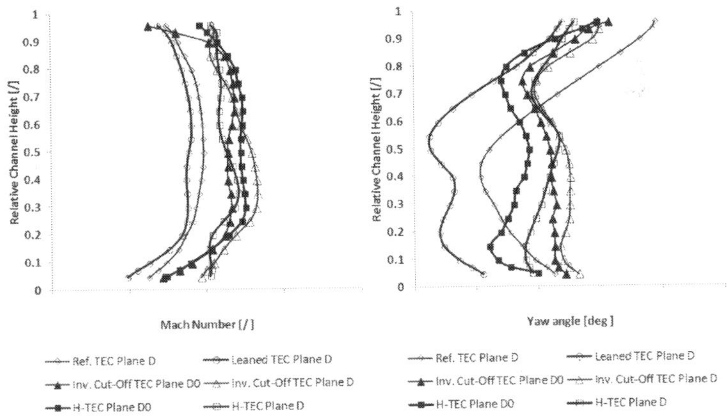

Figure 6. Mach number distribution (left) and yaw angle distribution (right) at TEC outlet.

4.3. Loss estimation

A rough estimation of the total pressure loss coefficient $\zeta = \frac{\bar{\bar{p}}_{t,C} - \bar{\bar{p}}_{t,D}}{\bar{\bar{p}}_{t,C} - p_{ex}}$ from plane C upstream of the TEGVs to plane D downstream of the TEGVs is done. The total pressure has been mass averaged by means of the five-hole probe data. **Table 7** compares the change of total pressure loss of the three configurations. The loss coefficients have been normalised with the total pressure loss coefficient ζ_{ref} of the reference configuration. Inverse cutoff, the leaned TEC as well as the H-TEC configuration show higher loss coefficients than the reference TEC. That means that all configurations produce higher losses for the acoustically important operating point approach and it seems that they are more sensitive at off design conditions. However, the difference between the leaned and the inverse cutoff TECs is very small and within the measurement uncertainty.

	Total pressure loss $\frac{\zeta}{\zeta_{ref}}$	
Configuration	C-D	C-D0
Ref. TEC	1.0	-
Leaned TEC	~1.6	-
Inverse cutoff TEC	~1.6	~1.6
H-TEC	~1.3	~0.9

Table 7. Total pressure loss.

Further, it can be seen that the losses increase from plane D0 to D for the H-TEC. The losses are nearly constant for the inverse cutoff TEC from plane D0 to D. However, pressure loss measurements at aero design point showed a significant loss reduction from plane C to D for both, the aerodynamically optimised H-TEC and the inverse cutoff TEC when compared with the reference TEC $\frac{\zeta}{\zeta_{ref}}=0.9$. A comparison between the leaned TEC and the reference TEC show similar losses so that $\frac{\zeta}{\zeta_{ref}}\approx 1$.

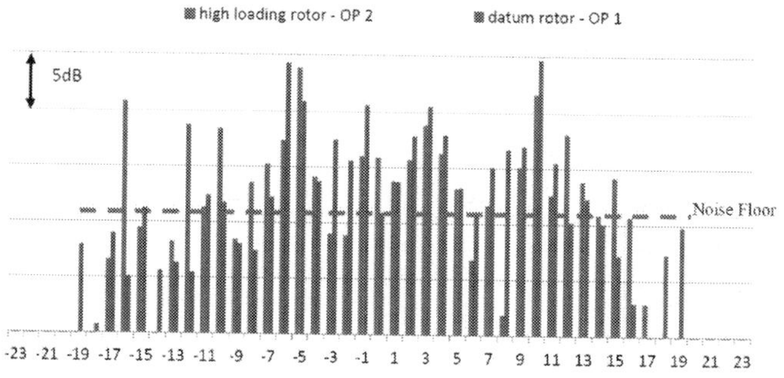

Figure 7. Comparison of modal PWL at LPT exit for two different stages; identical shaft power.

During the last decades a lot of institutions investigated the possibility to reduce engine weight by reducing the blade count of a low pressure rotor leading to highly loaded or even ultra-high-loaded turbine blades. Therefore, also the acoustical behaviour of such a rotor compared to a state-of-the-art design is investigated. At first, the results for the designated, acoustically relevant off-design point approach for the two rotors, called OP1 for the datum stage (higher rotational speed) and OP2 for the high loaded stage (lower rotational speed), are shown. Due to the change in loading of 20%, the resulting rotational speed for the two geometries differs by 10% (see **Table 1**). **Figure 7** shows a comparison of the sound power level of different azimuthal mode orders evaluated for the first blade passing frequency (BPF) for the two different rotor setups. The cutoff criteria for the two operating points shows that for operating point 1 (OP1), the number of propagating modes is 39 (±19) whereas for operating point 2 (OP2), the number of propagating modes is only 35 (±17). Positive modes rotate against the rotational direction of the rotor and negative modes rotate in the same direction as the rotor. The figure also clearly shows that modes dominating the sound field are different for the two rotor configurations.

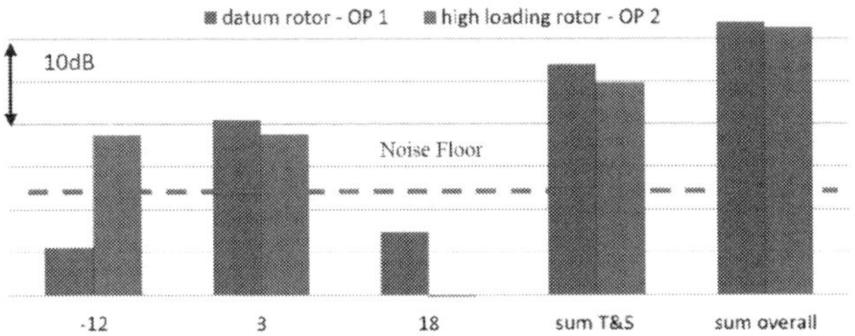

Figure 8. Comparison of modal PWL for propagating modes (rotor-TEGV-interaction).

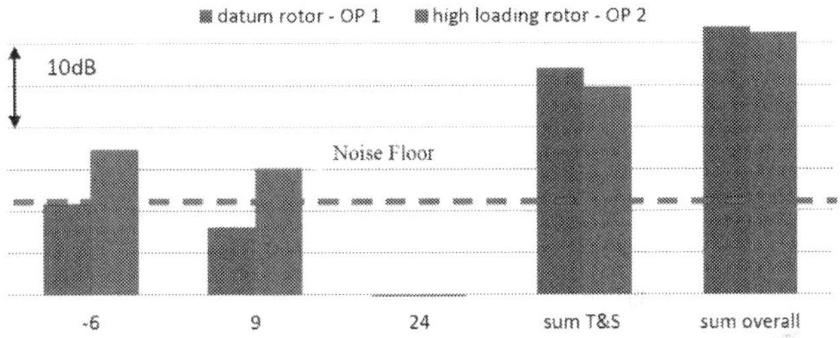

Figure 9. Comparison of modal PWL for propagating modes (rotor-stator-interaction).

The dominant modes for the OP1 with the datum rotor are the modes m=-6 and +10, towering the next highest modes by approximately 4 dB. While the mode -6 is predicted by Tyler and Sofrin, the mode m=+10 could not be calculated. When, due to the higher loading of the second stage, the rotational speed is lowered, the modal distribution of the resulting flow field changes. The dominant mode for OP2 is the mode -5, that is, 2 dB higher than the second highest mode m=+10. The overall sound power level of the two configurations is shown in **Figure 8**. It can be seen that the PWL of the high loaded rotor with reduced rotational speed is reduced by 0.7 dB. Considering only the propagating modes predicted according to the theory of Tyler and Sofrin, the high loaded rotor shows a 2.5 dB lower PWL. The difference in power level between propagating modes according to Tyler and Sofrin and considering all modes found in the measurements is approximately 5 dB. The highest modes found in the analysis (m=+10 for OP1 and m=-5 for OP2) of the measurement data are not attributable to

Tyler and Sofrin. It is assumed that non-Tyler Sofrin modes are increased for the highly loaded stage. Also, the sound power due to airfoil lift is assumed to be higher for the high loaded rotor. In addition, **Figure 8** depicts the modal PWLs of the Tyler-Sofrin modes of the interaction between the rotor and the TEGVs.

The modes according to Tyler and Sofrin are m=-12, m=+3, and m=+18. For the datum stage the highest amplitude for the rotor-TEGV interaction mode is for m=+3. For the high loaded stage both modes m=-12 and m=+3 show the same power level amplitude. Mode m=-18 is cut off for the high load stage. Its contribution to the overall power level for the datum stage is negligible. **Figure 9** depicts the modal sound power levels of the stator-rotor interaction. The main interaction mode would be the mode m=+24 that is cut off.

However, due to the small axial distance between stator and rotor the mode has not fully decayed and a scattering of that mode at the TEGVs is possible resulting in additional modes m=+9 and m=-6. The remaining interactions between the rotor and the inlet guide vane of the testrig are depicted in **Figure 10**.

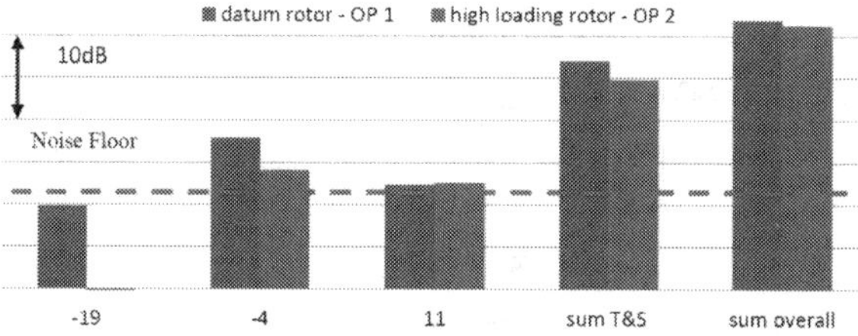

Figure 10. Comparison of modal PWL for propagating modes (rotor-IGV-interaction).

The mode m=-4 has the highest PWL for both configurations, but is still approximately 7 dB lower than the sum of the Tyler and Sofrin modes. The mode m=11 is significantly lower as well as the mode m=-19. This mode is only cuton and therefore able to propagate for the datum stage but is cut off and cannot propagate for the high loaded one. Comparing the different PWLs due to the interactions according to Taylor and Sofrin, the largest contributors to the overall PWL for the high loaded stage are interactions between the rotor and the TEGV as well as the mode m=-6, which is a rotor-stator interaction. For the datum stage the rotor-TEGV interaction (m=+3) and the rotor-IGV interaction (m=-4) are the main contributors to the overall sound power level.

4.4. TEGV pressure loss estimation

Also for that investigation a rough estimation of the total pressure loss coefficient from plane C upstream to plane D downstream of the TEGVs is done by means of the five-hole probe data. **Table 8** shows the total pressure loss coefficients between both configurations. They have been normalised with the total pressure loss coefficient ζ_{ref} of the reference configuration. The losses produced by the TEGV are slightly increased by about 16% for the high stage loading stage upstream the TEGV than for the datum stage. It seems that flow structures from the rotor are mixed out in that TEC region and that the high loaded rotor produces flow structures resulting in higher mixing losses.

	Total pressure loss $\frac{\zeta}{\zeta_{ref}}$
Configuration	**C-D**
Datum stage	1.0
High loading stage	~1.16

Table 8. Total pressure loss.

5. Research on TCFs

5.1. Frequency spectra analysis

The frequency spectra in **Figure 11** (left and right) report the comparison between the two setups C1 (black) and C3 (red). **Figure 11** (left) shows that the highest amplitudes may be identified at the blade passing frequency of the HP rotor and its first harmonic. The red graph is shifted by 100 Hz for a better visibility.

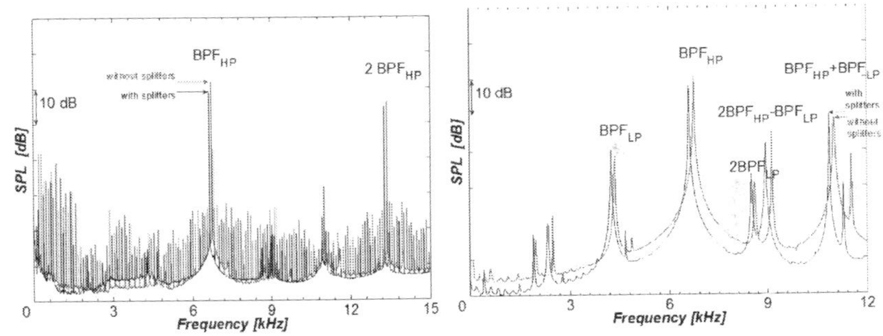

Figure 11. Frequency spectra of the time-signal reconstructed by the HP trigger for the setup without splitters (red) and with splitters (black). Right: Frequency spectra of the microphone array for the sum of the BPF of the two rotors after rotor synchronic averaging; without splitters (red) and with splitters (black).

An opposite trend is observed at the $BPF_{LP} + BPF_{HP}$, where the amplitude is 10 dB lower than the one of the HP rotor alone. The amplitude at the frequency $2BPF_{HP} - BPF_{LP}$ is almost the same as the logarithmic sum of the two rotors. Nearly 20 dB difference in the sound pressure level is observed between the amplitudes of the HP and the LP rotor. At BPF_{LP} the pattern of the peaks of the two different setups is like the one at $BPF_{LP} + BPF_{HP}$.

5.2. Azimuthal mode analysis BPF

An overview of the propagating modes is given in **Table 9** for the different blade passing frequencies. The modes are asymmetric, because the theoretical prediction taken into account for a swirl model [20].

Frequency	Range of the modes
BPF_{HP}	-44 : 8 : 36
$2BPF_{HP}$	-96 : 8 : 80
BPF_{LP}	-32 : 8 : 24
$2BPF_{LP}$	-64 : 8 : 48
$BPF_{LP} + BPF_{HP}$	-84 : 8 : 68

Table 9. Range for the propagating modes dependent on the BPF.

5.3. HP rotor noise

For explanation of the complex pattern of the sound field due to the vane/blade interactions, the modes predicted by Tyler and Sofrin at the BPF_{HP} are listed for the baseline setup (also given in Lengani et al. [21]): HP stator-HP rotor interaction creates modes $m=36+k\Delta 24=..-36;-12;\ 12;36$. HP rotor-LP stator interaction results in modes $m=36+k\Delta 16=..-28;-12;4;20;36$. The interaction of HP stator-HP rotor-LP stator creates modes $m=36+k_1\Delta 24+k_2\Delta 16=..-44;-36;-28;-20;\ -12;-4;\ 4;12;20;28$. These modes may be identified in **Figure 12** on the left side for the baseline configuration C1. The picture shows the sound pressure level (SPL) for the BPF_{HP} in dB, for the different azimuthal mode orders m. In **Figure 12** propagating modes are marked with black bars. It is clear that the propagating mode ranges from mode -44 to +36. Mode orders $m<-44$ and $m>+36$ are cut off; however, a low amplitude is visible; hence they are not fully decayed within the duct. Furthermore, all modes with high amplitudes can be predicted according to Eq. (7) as linear combination of HP vane-HP blade-LP stator, and they are scattered by 8 (see also **Table 9**). Hence, noise emanated from the HP rotor has to be attributed to its interaction with the up-and downstream vanes. For the splitter setup the prediction of modes for the HP-rotor-LP stator interaction as well as for the HP stator-HP rotor-LP stator interaction can be done with different vane numbers. The LP vane count may be considered equal to 48 because of the 32 additional splitters. However the geometry of the splitters is not the same as the one of the struts and the splitter leading edge is far downstream in the strut passage, but for a first estimation it can be assumed that the interaction is similar. Therefore, the theory of Tyler and Sofrin is not fully satisfied but can be applied. For the interaction of HP stator-HP rotor-LP vanes the following modes may be predicted in the ideal case of assumed 48 identical LP vanes: $m=36+k_1\Delta 24+k_2\Delta 48=..-36;\ -12;12;36;$ Without this assumption and further considering that

LP vanes and splitters as separate vane rows (HP stator-HP rotor-LP vane-splitter interaction) the modes can again be obtained from this simple linear combination:

$$36 + k_1 \cdot 24 + k_2 \cdot 32 + k_3 \cdot 16 = .. - 44; -36; \ -28; -20; \ -12; -4; 4; 12; 20; 28; 36$$

The modes propagating in the case of the splitter setup are shown on the right side of **Figure 12**. The modes, which change significantly between the two setups, are labelled in the picture. Particularly, the power levels of modes $m = -44$ and $m = -28$ are significantly reduced by almost 20 dB for the configuration with splitters. Also the amplitude of mode $m = +28$ is 10 dB lower. It should be noticed that these modes are not predicted assuming the ideal LP vane number of 48. It seems that the presence of splitter vanes changes the acoustic transmission of sound waves through the LP vanes for modes which are instead propagating for the baseline case. Further, modes from HP stator-HP rotor interaction are partially scattered at the leading edges of the 16 LP vanes and partially scattered further downstream in the LP vane passage at the 32 splitter leading edges. The modes mean value in the setup with the splitters is 5 dB lower than the one in the baseline setup.

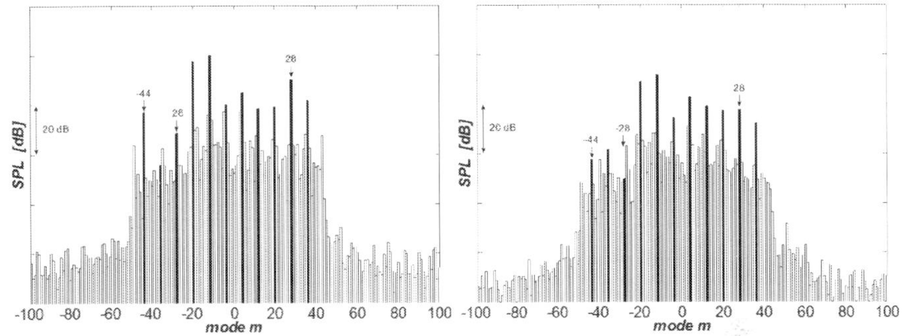

Figure 12. Azimuthal mode analysis of the flow with the HP trigger; without splitters (left) and with splitters (right).

5.4. LP rotor noise

Figure 13 depicts the sound power levels of the modal decomposition for the BPF_{LP}. This figure shows the baseline configuration on the left side and the splitter configuration on the right side. The interaction of the LP vane and the LP blade generates the following modes: $m = -72 + k\Delta 16 = .. -72; -56; -40; -24; -8; 8; 24$. These modes can be clearly identified in the left side of **Figure 13**, and are additionally indicated in black. The highest sound pressure level is observed for mode $m = -24$. In-between $-24 < m < 24$ other modes are blackened (e.g. $m = -16, \ 0, \ +16$) that can be predicted when the linear combination of LP vane and blade additionally considers the interaction with the HP vane ($-72 + k_1\Delta 16 + k_2\Delta 24$). However, the interaction between the HP vane and LP blade is weak (e.g. there is a difference of more than

20 dB between mode $m = -24$ and $m = -16$). Modes -72, -56, and -40 are still visible in the figure. These modes are generated by the LP rotor itself and by the LP stator-LP rotor interaction. However, those modes are cut off and therefore have a lower amplitude than the cuton modes. The cutoff modes will decay further downstream of the duct. In the case of the splitter setup (right side of **Figure 13**), again a different LP vane count may be considered. Ideally, the LP vane-LP rotor interaction would consist of the modes: $m = -72 + k\Delta 48 = ..-72; -24; 24$. Mode $m = -24$ has still the highest amplitude; however it is now in the same order of magnitude of the amplitude of mode $m = -8$. In the assumed ideal case of the splitter setup, mode $m = -8$ should not be there. However, as observed by Spataro et al. [22, 23] the unsteady effects induced by strut and splitter vane wakes differ considerably downstream of the LP rotor. The modes generated due to the interaction of the 48 LP vanes with the LP blades may be predicted when the modes are scattered at the 16 struts (e.g. mode $m = -8$ is obtained adding the 16 vanes to mode -24). In the case of the splitter setup, the mode $m = -8$ is more than 20 dB higher than in the baseline case. Also the sound pressure level of the modes generated due to the interaction HP vanes-LP vanes-LP-rotor is altered. It seems that the splitter design shifts the important modes to lower orders (e.g. mode -8 which was hardly identifiable for the baseline setup). As can be seen in the mean spectrum (**Figure 11** (right)), the modes mean value is almost 3 dB lower for the baseline configuration.

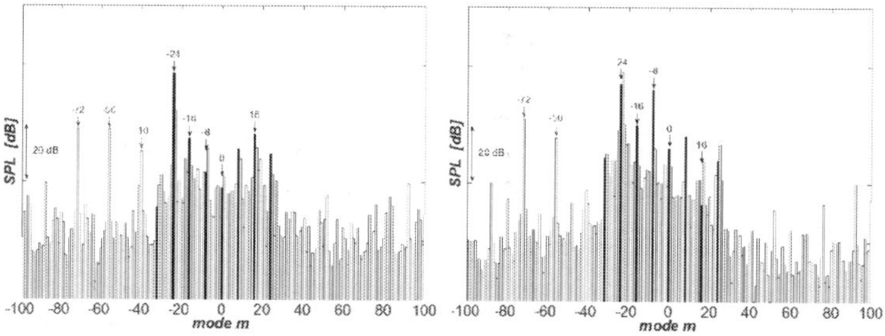

Figure 13. Azimuthal mode analysis of the flow with the LP trigger; without splitters (left) and with splitters (right).

Also for the splitter setup the modes $m = -72$, -56 are cut off as previously discussed for the baseline design. Mode $m = -40$ is significantly reduced by more than 20 dB in the splitter configuration. The high sound pressure level at the BPF_{LP} may be due to the effect of "shifting" modes towards lower orders and not necessarily to an enhanced unsteady interaction. Unsteady measurements by means of a fast response aerodynamic pressure probe downstream of the LP rotor reported in Spataro et al. [23] revealed that the unsteady pressure fluctuations, evaluated for the LP rotor phase, are of comparable order of magnitude for both setups.

In **Figure 14** the sound power levels of the modal decomposition is shown for the sum of the blade passing frequencies of the two rotors ($BPF_{LP} + BPF_{HP}$) for the baseline setup (left) and the setup with splitters (right). The modes mean value is almost the same for both setups.

However, there are some predominant modes clearly visible. Its amplitudes changing up to 23 dB and results from stator/rotor/stator/rotor interaction. Although mode $m = -28$ has one of the highest sound power levels for the baseline setup, its amplitude is significantly reduced for the splitter design. But in the case of the splitters, other modes like $m = -20$ and $m = -4$ appear, with amplitudes that are 20 dB and 14 dB, respectively, higher than in the baseline setup. Therefore the sound pressure level of both setups is almost the same for that specific frequency (sum of the blade passing frequencies of the two rotors). The largest reduction of the modes mean amplitude for the splitter setup is associated to the HP rotor. There is a 5 dB lower amplitude at the BPF_{HP} compared to the baseline design. At $2BPF_{HP}$ the results and trends of the azimuthal mode analysis are similar. A sound pressure level reduction of 4 dB was achieved with the splitter design. In contrast, the sound pressure level at BPF_{LP} and $2BPF_{LP}$ is 3 dB lower for the setup without splitters, and at the $BPF_{LP} + BPF_{HP}$ it is nearly the same in both cases.

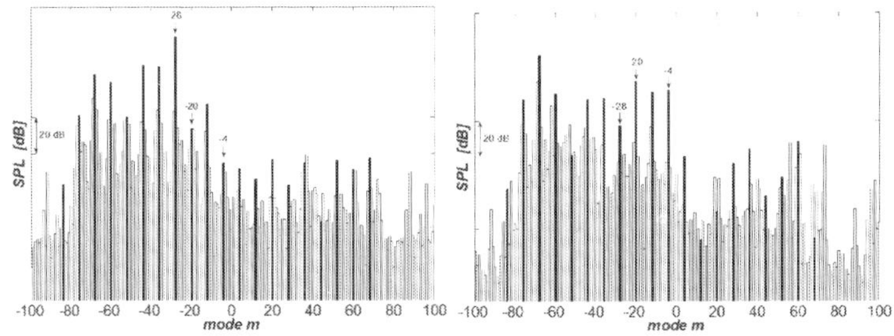

Figure 14. Azimuthal mode analysis of the flow with the two triggers; without splitters (left) and with splitters (right).

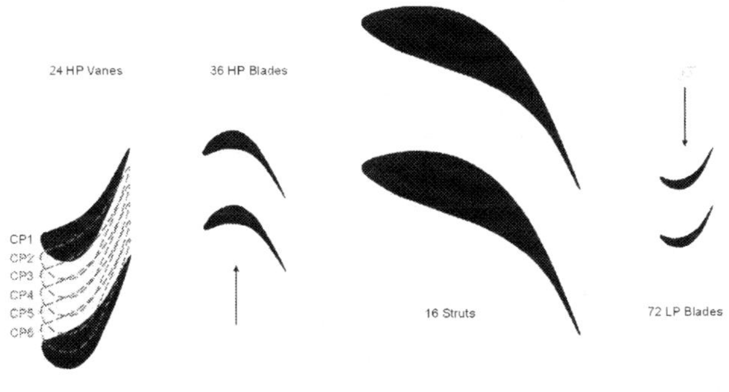

Figure 15. Section of the HP an LP stage showing the different clocking positions.

The comparison of the baseline setup C1 and the shortened setup C2 showed an increase of the sound pressure level between 5 and 9 dB dependent on the operating point. Especially the interaction modes between the struts and the LP rotor increase due to the 10% shortening of the duct length.

As a next step the influence of airfoil clocking on the acoustics was investigated. The results of this investigation of noise generation and propagation for different clocking positions (CP) of the HP vanes and the struts are presented. A meridional section presenting the six vane-vane positions can be found in **Figure 15**.

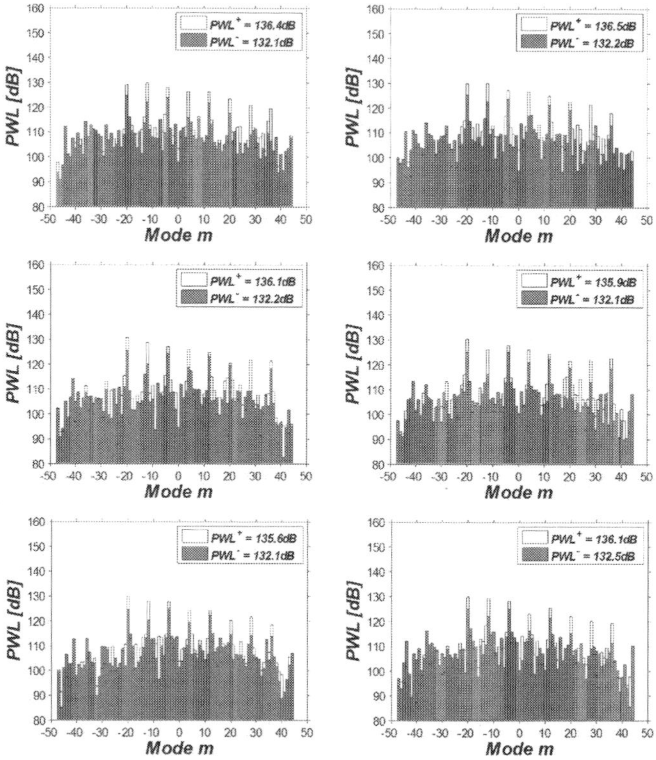

Figure 16. Radial mode analysis with reference to the HP trigger.

The sound power levels for BPF_{HP} both in positive (gray bars in **Figure 16**) and in negative (white bars in **Figure 16**) flow direction were calculated and plotted for all the radial modes over the azimuthal modes at the abscissa. Furthermore an overall sound power level was determined by logarithmic addition for both directions of propagation. Further, **Figure 16** shows the sound power level in decibel (dB) over the propagatable azimuthal modes m summed over the radial mode order n in the up- and downstream direction for all clocking positions (CP1–CP6).

Several interaction modes are clearly visible in the figures. The amplitudes of these significant modes are 20 dB larger than those of the non-interaction modes. In particular, the modes $m = -20$ and $m = -12$ are dominant for all clocking positions except CP4 here the mode $m = -4$ is higher than $m = -12$. Clocking position CP4 is that relative stator-stator position, where the sum of the most significant modes has its minimum. For the following discussion only the six dominating modes with the highest sound power levels are considered. With this analysis of certain modes the origin of a higher or a lower overall sound power level dependent on the clocking position can be determined. The most significant modes, which can be derived from different stator/rotor/stator interactions that can be predicted with Eq. (6) or Eq. (7), are selected and compared for all the different clocking positions. These interaction modes have the most influence on the overall sound power level, because if the difference between two incoherent sound signals is larger than 10 dB, the acoustic source with the smaller level has no noticeable influence on the sum of the power levels. The following significant azimuthal interaction modes (HP stator, HP rotor, and TMTF) were predicted: $m = 36 + k_1 \Delta 24 + k_2 \Delta 16 = .. -20; -12; -4; 4; 12; 20; \ldots$. Mode $m = -20$ has the highest amplitude. This mode rises over the first three clocking positions (see **Figure 17**) reaching its maximum at clocking position 3, which is close to the optimum aerodynamic clocking position.

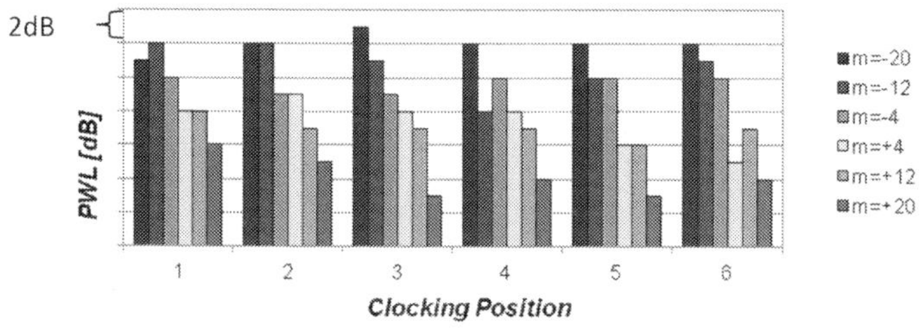

Figure 17. Six most dominant modes at seven different clocking positions.

Defining the optimum aerodynamic clocking position was performed by means of pre-test CFD calculations. In that way, most of the wakes of the HP stator impinge on the leading edges of the TMTF struts [24]. It is important that mode $m = -20$ can only be predicted by the interactions of HP stator, HP rotor and the TMTF. It is assumed that this particular mode originates from an effect where the TMTF plays a significant role. That means that the wakes of the HP stator impinge on the leading edges of the struts and the flow through the strut passage remains more or less undisturbed. However, the flow downstream the turning mid turbine frame shows larger differences of flow quantities between the wake and the main passage flow. By changing the relative vane-vane position (=clocking also known as stator indexing) to the fourth clocking position the amplitude of $m = -20$ is reduced. For the next two clocking positions 5 and 6 the sound power level remains almost the same. The high sound power level of mode $m = -20$ is due to the vane/rotor/strut interaction. However, this mode is neither predicted by the HP stator-HP rotor interaction nor by the HP rotor/TMTF interaction. The unsteady interaction of the HP-stage is scattered by the downstream turning struts in the flow path. Mode $m = -20$ is generated by the scattering of the HP-stage interaction at the TMTF. **Figure 17** shows that the sound power levels of mode $m = -12$ decrease from clocking position 1 to 4 and increase again from clocking position 4 to 6. The minimum sound power level can be seen at clocking position 4. The mode $m = -12$ is either generated by the HP stator-HP rotor interaction or by the HP rotor-TMTF interaction. In case of mode $m = -12$ the strongest influence of the different clocking positions can be determined. This particular mode is reduced by 4 dB when changing the relative vane-vane position from CP1 to CP4. For the modes $m = +12$ and $m = +20$ a similar trend to mode $m = -12$ can be observed. These modes also show significant changes of the sound power levels of up to 3 dB due to different relative positions of the HP vanes and the TMTF struts. Both modes reach their minimum sound power level at clocking position 5, whereas the level of the amplitudes is decreasing from 1 to 4. The mode $m = +12$ is generated by the interaction of the HP vanes and the HP blades but also by scattering of the HP stage interaction modes at the TMTF. The mode $m = +20$ is always generated in conjunction with the TMTF-vanes, either with the HP rotor or with the HP stage. While the amplitude of mode $m = -4$ seems to be almost constant for all clocking positions, mode $m = +4$ changes from CP2 to CP6 by 4 dB, whereas at the last clocking position the amplitude of m=+4 has its lowest value. Both modes are the result of the interaction of the HP-stage and the TMTF. Summing up the sound power levels (depicted in **Figure 17**) reveals that there is a minimum sound power at clocking position CP4. A difference in sound power level of app. 2 dB between the acoustically best (CP4) and worst (CP2) clocking position is observed.

6. Conclusions

Three different turbine exit casings with different turbine exit guide vane (TEGV) designs have been compared to a state-of-the-art (reference) TEGV design. The possible reduction of sound power levels when applying the different designs and a rough estimation of the aerodynamic losses have been presented. When comparing the overall reduction of PWL (only considering the main airfoil interaction modes) it was revealed that the acoustically optimised inverse cutoff TEC has the largest reduction of sound power level of 14 dB. The aerodynamically optimised H-TEC even increases the overall sound power level by about 2 dB. The leaned TEC also decreases the PWL by about 11 dB, but is still as twice as loud as the inverse cutoff TEC. However, for the operating point approach the aerodynamic losses are increased for all TEC designs. The losses measured at the aero design point are lower for the aerodynamically optimised H-TEC and the inverse cutoff TEC than for the reference TEC. Both, the H-TEC and the inverse cutoff TEC provide a much more uniform yaw angle distribution downstream of the trailing edge. These results give confidence that it is possible to design an aerodynamically and acoustically optimised TEC. Further, the effect of a change in stage design onto a TEGV with a compound lean was investigated and presented in this chapter. It was shown that while keeping the shaft power constant, the noise emissions downstream of the TEGV can be reduced by 0.7 dB for the given geometry and operating point. This decrease in sound power level is not caused by a decrease in interactions, but rather by acoustic modes not attributable to these interactions. The overall sound power level of the modes is increased by 0.3 dB. When having

http://dx.doi.org/10.5772/62431

the same rotational speed for the high loaded stage and the datum stage, the main interaction modes are identical. In addition to that main modes, several additional modes are found, that have been leading to an overall increase in sound power level. The TEGV inlet flow show only minor differences between the two stage designs. The main change occurs in the tip region. The main flow features of the TEGV exit flow remain identical. But it was observed that the shape of the wake changes due to the changes in the rotor tip leakage flow upstream of the TEGV. This increases the total pressure loss of the TEGV by approximately 16%.

Because of the possible significant reduction of the generated noise at the turbine exit casing noise from more upstream engine components such as the turbine centre frame can become a problem. Therefore, the noise generation between high pressure turbine, turbine centre frame, and low pressure turbine was investigated. Three different setups have been compared to each other. A baseline case C1 is designed with turning struts. A second design C2 reduces the axial length of the turbine centre frame. The third design C3 is characterized by the presence of two non-lifting splitters embedded into the strut passage. In the frequency spectra, the peak at the BPF_{HP} is 3 dB lower for the splitter setup. In the azimuthal mode analysis the difference is even higher. The sound power level decreased 5 dB at the BPF_{HP} for the setup with the splitters. Additionally, both the frequency spectra and the azimuthal mode analysis have shown that the noise generated by the LP rotor is slightly higher for the setup with the splitters. The splitter design is suppressing some modes, while others are scattered or are even more pronounced than in the baseline configuration. The splitters reduce the overall noise propagation by 5 dB acting as a cutoff filter for the HP stage rotor. The comparison of the baseline setup C1 and the shortened setup C2 showed an increase of the sound pressure level between 5 and 9 dB dependent on the operating point. Additionally, an experimental investigation in order to explore the potential of different vane-vane clocking positions on the noise generation and propagation was conducted. The six most significant modes have been analyzed regarding their sound power levels for the six clocking positions. The modes varied in their absolute values for the sound power and also their trend over the clocking positions changed. An optimum clocking position for acoustics was found, but it does not to coincide with the optimum aerodynamic clocking position. The difference of the overall sound power level for the six most relevant modes between the optimum acoustic clocking position (CP4) and the aerodynamic one (CP3) is about 2 dB.

7. Nomenclature

Abbreviations

ACARE	Advisory Council for Aeronautics Research in Europe
AMA	Azimuthal mode analysis
BPF	Blade passing frequency

C1, C2, C3	Configuration 1, 2, 3
CP1,...,CP6	Clocking position 1 to 6
EGV	Exit guide vanes
HP	High pressure
HPT	High pressure turbine
IGV	Inlet guide vanes
LP	Low pressure
LPT	Low pressure turbine
OP1, OP2	Operating point 1, 2
PWL	Sound power level
RMA	Radial mode analysis
SPL	Sound pressure level
STTF-AAAI	Subsonic test turbine facility for aerodynamic, acoustic, and aeroelastic investigations
TCF	Turbine centre frame
TEC	Turbine exit casing
TEGV	Turbine exit guide vanes
TMTF	Turning mid turbine frame
TTTF	Transonic test turbine facility
T&S	Tyler and Sofrin modes

Symbols

A	Complex amplitude
B	Number of blades
c	Velocity, speed of sound
F	Normalisation factor
f	Modal shape factor
h	Channel height
h	Harmonic index, integer number
i	Imaginary unit
J	Bessel function of first kind
k	Wave number
k	Integer number
Ma	Mach number
m	Azimuthal mode order
P	Sound power

http://dx.doi.org/10.5772/62431

p	Pressure
Q	Eigenvalue of the Bessel function
Re	Reynolds number
r	Radius
R	Outer radius
t	Time
V	Number of vanes
Y	Bessel function of second kind
α	Yaw angle
α	Axial wave number considering cuton frequency
α	Circumferential position
ω	Angular frequency
Ω	Angular frequency of swirl
σ	Eigenvalues of the Bessel function
ϱ	Density (air)
ζ	Pressure loss coefficient
Δ	Laplace operator, delta
<>	Deterministic periodic component

Index

1,2,3,..	Stage count
ax	axial
C, D	Measurement planes
ex	Exhaust casing
H-TEC	High lift design TEC
inv.	Inverse cut-off TEC
leaned	Leaned TEC
m	Azimuthal mode order
n	Radial mode order
ref	Reference
t	Total

Superscripts

$\circledcirc^{-}$	Time averaged properties
$\circledcirc^{\sim}$	Ensemble averaged properties, modified wave number
$\circledcirc'$	Stochastic fluctuations

$\bar{\circ}$	Mass averaged
+	Propagation in flow direction
-	Propagation against flow direction

Author details

Andreas Marn[1*], Christian Faustmann[2] and Thorsten Selic[3]

*Address all correspondence to: andreas.marn@tugraz.at

1 Institute for Thermal Turbomachinery and Machine Dynamics, Graz University of Technology, Graz, Austria

2 Audi AG, Ingolstadt, Germany

3 Elin Motoren GmbH, Preding/Weiz, Austria

References

[1] D. Sutliff, Rotating Turbofan Duct Mode Measurment System, "Proceedings of 9th National Conference on Noise Control Engineering", October 2005.

[2] Broszat D, Kennepohl F, Tapken U, Moser M, Heitmeir F. Validation of an Acoustically 3D-Designed Turbine Exit Guide Vane. In: Proceedings of the 16th AIAA/CEAS Aeroacoustics Conference, Stockholm, Sweden; June 8-9; 2010. DOI: 10.2514/MAERO10

[3] Broszat D, Selic T, Marn A. Verification of the Inverse Cut-Off Effect in a Turbomachinery Stage. Part 1 – Numerical Results. In: Proceedings of the 18th AIAA/CEAS Aeroacoustic Conference, Colorado Springs, Colorado, USA; June 4-6; 2012. DOI: 10.2514/MAERO12

[4] Broszat D, Selic T, Marn A. Verification of the Inverse Cut-Off Effect in a Turbomachinery Stage. Part 2 – Comparison to Experimental Results. In: Proceedings of the 19th AIAA/CEAS Aeroacoustic Conference, Berlin, Germany; May 27–29; 2013. DOI: 10.2514/MAERO13

[5] Tyler J M, Sofrin T G. Axial Flow Compressor Noise Studies. SAE Transactions. 1962;70:309-332.

[6] Moser M, Tapken U, Enghardt L, Neuhaus L. An Investigation of Low Pressure Turbine Blade-Vane Interaction Noise: Measurements in a 1.5-Stage Rig. Proceedings of the

Institution of Mechanical Engineers, Part A: Journal of Power and Energy. 2009;223(6): 687-695. DOI: 10.1243/09576509JPE823

[7] Faustmann C, Lengani D, Spataro R, Marn A, Göttlich E, Heitmeir F. Experimental Investigation of the Noise Generation and Propagation for Different Turning Mid Turbine Frame Setups in a Two-Stage Two-Spool Test Turbine. In: Proceedings of the ASME Turbo Expo, San Antonio, Texas, United States of America; June 3–7; 2013. DOI: 10.1115/GT2013-95698

[8] Hussain W, Reynlds A. The Mechanics of an Organized Wave in Turbulent Shear Flow. Journal of Fluid Mechanics. 1970;41:241-258. DOI: 10.1017/S0022112070000605

[9] Lengani D, Santner C, Spataro R, Göttlich E. Analysis Tools for the Unsteady Interactions on a Counter-Rotating Two-Spool Turbine Rig. Experimental Thermal Fluid Science. 2012;42:248-257. DOI: 10.1016/j.expthermflusci.2012.05.010

[10] Enghardt L, Neuhaus L, Lowis C. Broadband Sound Power Determination in Flow Ducts. In: Proceedings of the 10th AIAA/CEAS Aeroacoustic Conference, Manchester, United Kingdom; May 10–12; 2004. DOI: 10.2514/MAERO04

[11] Enghardt L, Tapken U, Kornow O, Kennepohl F. Acoustic Mode Decomposition of Compressor Noise Under Consideration of Radial Flow Profiles. In: Proceedings of the 11th AIAA/CEAS Aeroacoustic Conference, Monterey, California, United States of America; May 23-25; 2005. DOI: 10.2514/MAERO05

[12] Enghardt L, Tapken U, Neise W, Kennepohl F, Heinig K. Turbine Blade/Vane Interaction Noise: Acoustic Mode Analysis Using In-Duct Sensor Arrays. In: Proceedings of the 7th AIAA/CEAS Aeroacoustic Conference, Maastricht, Netherlands; May 28–30; 2001. DOI: 10.2514/MAERO01

[13] Holste F, Neise W. Noise Source Identification in a Propfan Model by Means of Acoustical Near Field Measurements. Journal of Sound and Vibration. 1997;203(4):641-665. DOI: 10.1006/jsvi.1996.0890

[14] Laguna J D, Barteld M, Seume J. Impact of Swirl on the Sensitivity of the Radial Mode Analysis in Turbomachines. In: Proceedings of the ASME Turbo Expo, San Antonio, Texas, United States of America; June 3–7; 2013. DOI: 10.1115/GT2013-95460

[15] Tapken U, Enghardt L. Optimization of Sensor Arrays for Radial Mode Analysis in Flow Ducts. In: Proceedings of the 12th AIAA/CEAS Aeroacoustic Conference, Cambridge, Massachusetts, United States of America; May 8–10; 2006. DOI: 10.2514/MAERO06

[16] D. Sutliff, „Rotating Turbofan Duct Mode Measurment System," *Proceedings of 9th National Conference on Noise Control Engineering*, Minneapolis, Minnesota, United States of America; October 2005.

[17] Morfey C. Sound Transmission and Generation in Ducts with Flow. Journal of Sound and Vibration. 1971;14(1):37-55. DOI: 10.1016/0022-460X(71)90506-2

[18] Enghardt L, Zhang Y, Neise W. Experimental Verification of a Radial Mode Analysis Technique Using Wall-Flush Mounted Sensors. Journal of The Acoustical Society of America. 1999;105(2):1186–1187. DOI: 10.1121/1.425598

[19] Sijtsma P, Zillmann J. In-Duct and Far-Field Mode Detection Techniques. In: Proceedings of the 13th AIAA/CEAS Aeroacoustic Conference, Rome, Italy; May 21-23; 2007. DOI: 10.2514/MAERO07

[20] Taddei F, De Lucia M, Cinelli C, Schipani C. Experimental Investigation of Low Pressure Turbine Noise: Radial Mode Analysis for Swirling Flows. In: Proceedings of 12th International Symposium on Unsteady Aerodynamics, Aeroacoustics & Aeroelasticity of Turbomachines, London, United Kingdom; 2009.

[21] Lengani D, Santner C, Spataro R, Paradiso B, Göttlich E. Experimental Investigation of the Unsteady Flow Field Downstream of a Counter-Rotating Two-Spool Turbine Rig. In: Proceedings of the ASME Turbo Expo, Copenhagen, Denmark; June 11-15; 2012. DOI: 10.1115/GT2012-68583

[22] Spataro R, Göttlich E, Lengani D, Faustmann C, Heitmeir F. Development of a Turning Mid Turbine Frame with Embedded Design-Part I: Design and Steady Measurements. In: Proceedings of the ASME Turbo Expo, San Antonio, Texas, United States of America; June 3-7; 2013. DOI: 10.1115/GT2013-95279

[23] Spataro R, Göttlich E, Lengani D, Faustmann C, Heitmeir F. Development of a Turning Mid Turbine Frame with Embedded Design-Part II: Unsteady Measurments. In: Proceedings of the ASME Turbo Expo, San Antonio, Texas, United States of America; June 3-7; 2013. DOI: 10.1115/GT2013-95280

[24] Schennach O, Pecnik R, Paradiso B, Göttlich E, Marn A, Woisetschläger J. The Effect of Vane Clocking on the Unsteady Flow Field in a One-and-a-Half Stage Transonic Turbine. Journal of Turbomachinery. 2008;130:031022-1 – 031022-8. DOI: 10.1115/1.2777199

Chapter 2

Lubricants for Turbine Engines

David W. Johnson

Additional information is available at the end of the chapter

http://dx.doi.org/10.5772/62394

Abstract

The lubricant systems used in turbine engine applications are discussed with respect to the particular problems associated with aircraft applications. After initially describing the relevant specifications, the typical basestocks are described along with some common degradation schemes. The additive systems, including antioxidants, anti-foaming agents, and anti-wear additives needed to achieve the typical specifications, are described along with their mechanism of action and degradation mechanisms. The methods used for the monitoring of lubricant health, including in-line and offline methods, are also discussed. Finally, future changes in specifications, basestocks, and additives are discussed with respect to new, high-performance bearing materials.

Keywords: synthetic lubricants, anti-oxidants, phosphate esters, condition monitoring, tribology, lubricant additives

1. Introduction

Lubrication is critical in any application where moving parts are involved. Aircraft involve large numbers of different types of moving parts, many of which are associated with the propulsion system. The objectives of lubrication are primarily to improve energy efficiency by reducing friction and to improve component lifetime by reducing wear of the moving parts. An added benefit of an efficient lubrication system is a reduction in the engine operating temperature because of the heat dissipation through the lubricant system [1].

Turbine engine lubrication is an extremely demanding application for lubricants for several important reasons, as follows:

- Reliability due to the consequences of failure. Aircraft applications are the only applications where it is not possible to stop immediately and investigate a lubricant failure indication.

- High operating cost associated with added weight, making backup and redundant systems highly undesirable. This requirement leads to small volumes of lubricant and results in high heat dissipation requirements and high operating temperatures.
- Wide range of temperatures, pressures, and speeds that the lubricant is exposed to under normal operating and storage conditions.

The combination of these requirements eliminates the possibility of an ideal natural lubricant or synthetic lubricant meeting all of the requirements. As speeds and temperatures increased, automotive lubricants initially used in aircraft were replaced by specialty lubricants for aircraft engines. As the propulsion systems changed from conventional engines to turbine engines, lubricants based on mineral oils were inadequate and a new class of lubricants was needed [2].

In the years after World War II, turbine engines were developed and began to dominate both military and commercial aircraft propulsion. Turbine engines require the lubricant to be stable and fluid over a wide range of temperatures. Ground starting temperature may be as low as -54°C and temperature as high as 300°C may be observed at times in the operating engine. Over time, a number of different lubricants were developed with very different chemistries and compatibilities. Many different locations had lubricants that were developed to suit the local conditions and needs. With a number of incompatible fluids, the global nature of aircraft use, and the grave consequences of mistakes, a need for standardization became apparent. To address this problem, a system of lubricant specifications has been developed for turbine engine lubricants.

2. Early basestocks

The earliest lubricants used in turbine engines were highly refined mineral oils. Mineral oils, however, lack the high temperature stability to withstand the high temperatures found in different points of the engine. Severe oxidation of the light mineral oils results in significant increases in the viscosity of the oil. The problems of severe oxidation and degradation highlighted the need for a new class of oils, which lead to the development of a class of synthetic aviation turbine oils [3].

The search for a new class of synthetic aviation lubricants led to the examination of organic esters as basestocks. Over 3500 different esters were examined for lubricant use between 1937 and 1944. A number of candidates had some of the required properties, including thermal stability, pour point, or boiling point, but none met all of the requirements. It should be possible based on the number of di-acids and alcohols available to achieve the desired viscosity and other properties with a pure compound; however, to achieve the wide liquid range required for Air Force needs, the blending of multiple esters was thought to be preferable to a single compound. The U.S. pour point of −40°C and high temperature requirement for a 3 cS fluid could be met by the reaction of simple mono-alcohols with dibasic carboxylic acids. The earliest commercial lubricant based on organic esters involved branched alcohols such as 2-ethylhexyl alcohol and adipic acid, azelic acid, and sebacic acid (**Figure 1**).

http://dx.doi.org/10.5772/62394

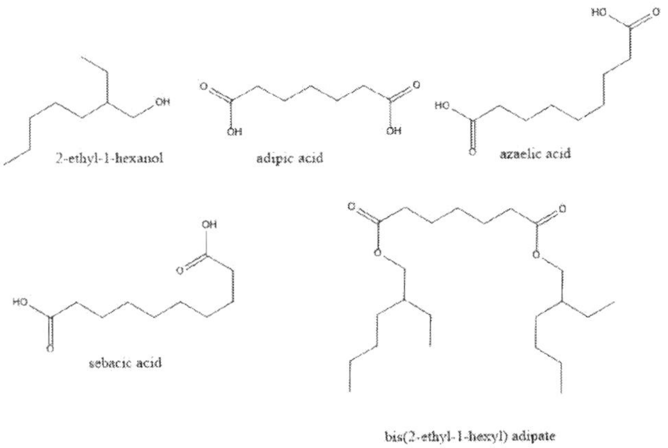

Figure 1. Structures of common components of lubricant esters.

Higher viscosity requirements of the British could be met using longer-chain acids and alcohols, but the low temperature requirements could not be met by these combinations. The solution was to use a slightly more viscous blend of the 3cS fluid with a viscosity improver, typically a polyglycol.

Two fundamentally different and incompatible lubricants were both unacceptable and major potential hazards. The use of the wrong lubricant was considered unacceptable even for emergency use. In the 1960s, an intermediate viscosity oil was developed with a viscosity of 5.5 cS at 99°C that could replace both of the fluids, especially for future engines. This fluid also was adopted by the U.S. Navy for use in its aircraft. The advent of supersonic aircraft led to a need for even higher oil temperatures and improved stability.

Hydrolysis is a characteristic reaction of esters in the presence of water. Hydrolysis can be avoided through the use of longer-chain and branched alcohols. The primary route for the degradation of the esters used in the lubricants described above is through an elimination reaction initiated by the loss of the hydrogen atom on the β carbon to the carboxylate (**Figure 2**).

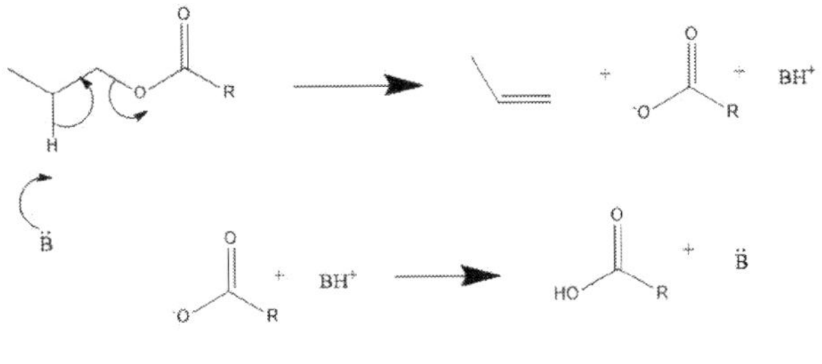

Figure 2. Mechanism for β-elimination.

The β-elimination results in the decomposition of the ester to alkenes, carboxylic acids. Although the mechanism involves a base, it is simply an acceptor for the proton and could be a variety of species present in the lubricant. Because β-elimination requires a hydrogen atom on the β carbon atom, the elimination of hydrogen atoms at that position resulted in a series of more thermally stable lubricants.

3. Current basestocks

As turbine engine technology developed and the desire for supersonic flight became the norm for the military, higher bearing load and bearing temperatures were needed. These developments required more stable lubricant basestocks. Because β-elimination reactions were one mode of degradation, alcohols that lacked hydrogen atoms at the β position were desired. Some common alcohols that fit the bill are the neopentyl polyols. These highly hindered alcohols react with acids of various chain lengths to form esters under acid catalyzed conditions, providing the lubricant esters in good yields. Current lubricants are based on the neopentyl polyols such as neopentyl glycol, pentaerythiol, and dipentaerythritol (**Figure 3**).

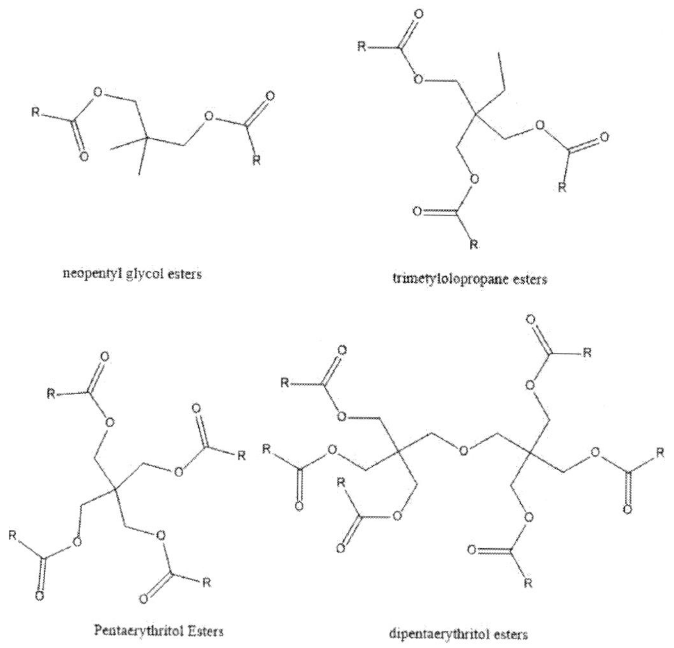

Figure 3. Structure of some esters commonly used in aircraft lubricants.

The esters used in the current aircraft lubricants use a mixture of carboxylic acids with between 5 and 9 carbon atoms, typically and may involve either linear acids or branched acids. Many common lubricants are based on a mixture of valeric acid, iso-valeric acid, heptanoic acid, nonanoic acid, and 3,5,5-trimethylhexanoic acid. The lubricants are prepared either from the acid-catalyzed esterification of the polyol with a mixture of the desired acids or it may involve the trans-esterification of a fatty acid methyl ester (FAME) with the polyol. The reaction can occur in a random manner, or can be controlled to occur stepwise, allowing more sterically hindered acids to react first and then complete the esterification with the less hindered acid [5]. The latter method allows esters to be prepared from biologically derived oils, such as vegetable oil [6]. The composition of the mixture of acids can be used to adjust the physical properties, including the viscosity and viscosity index of the resulting basestock [7]. The fact that a mixture is used also results in a wider liquid range for the resulting lubricant.

In general, esters based on the neopentyl polyols have excellent thermal stability and can perform well as lubricants. If, however, lower-quality reactants are used to prepare the ester, a substantially more reactive lubricant is obtained. Certainly, impurities in the alcohols that contain β-hydrogen atoms would significantly alter the high temperature stability. Other possible impurities include significant quantities of excess acid, which would increase the corrosivity of the oil; water, which has a significant solubility in the oil, can result in hydrolysis; and higher molecular weight polyols, such as dipentaerythritol or tripentaerythritol, can have a negative effect on the low temperature properties of the mixture.

The two grades of MIL-PRF-7808 that are available differ substantially in composition as is needed to obtain the difference in viscosity. The grade 3 lubricant because of the low temperature requirement has a higher content of low molecular weight acids and neopentylglycol as the dominant polyol. The total ion chromatogram from the gas chromatography-mass spectrometry (GC-MS) evaluation of a 3cS MIL-PRF-7808K grade 3 lubricant is shown in **Figure 4**.

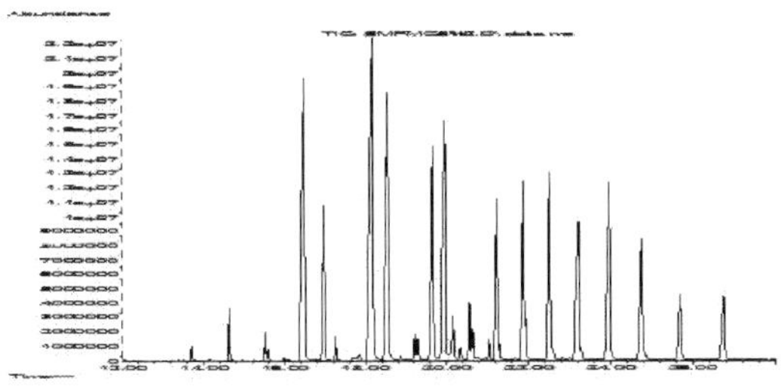

Figure 4. A portion of the total ion chromatogram of a MIL-PRF-7808K grade 3 lubricant.

In an effort to obtain superior high temperature properties to the MIL-PRF-7808K grade 3 and still retain better low temperature properties than is available with MIL-PRF-23699, MIL-PRF-7808K grade 4 was developed. The grade 4 oil used a different mixture of acids and alcohols to achieve a good compromise in properties. The total ion chromatogram from the GC-MS evaluation of a 4cS MIL-PRF-7808K grade 4 lubricant is shown in **Figure 5**.

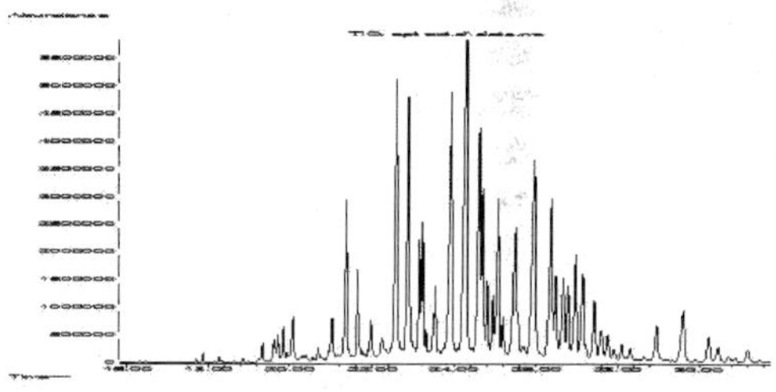

Figure 5. A portion of the total ion chromatogram of a MIL-PRF-7808K grade 4 lubricant.

This lubricant contains primarily the pentaerythritol esters of valeric acid, heptanoic acid, octanoic acid, and 3,3-dimetylheptanoic acid. Also observed between 20 and 23 min retention time are four peaks that correspond to isomers of the tri-cresyl phosphate additive. The large number of different esters helps with the wide liquid range desired, whereas the higher molecular weights lead to an increase in viscosity from 3 to 4 cS.

A further increase in molecular weight is desirable for MIL-PRF-23699 lubricants, which have a still better thermal stability and viscosity. The change in molecular weight can be seen in the distribution of esters in **Figure 6**.

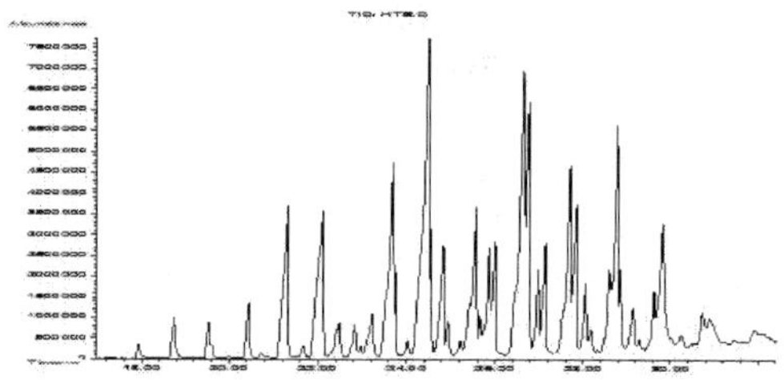

Figure 6. Total ion chromatogram of an MIL-PRF-23699 HTS lubricant showing the retention times where the esters elute.

The analysis of the acid composition of this lubricant indicates that at least six different acids are present in the mixture. The ability of multiple manufacturers to use somewhat different acid and alcohol blends yet achieve the same properties and have a compatible lubricant is thought to be a huge advantage.

Lubricant degradation at high temperatures is a significant problem. Degradation can result in an increase in acid number, an increase in viscosity, or an increase in the reactivity of the lubricant. For hindered esters where the β-elimination (discussed earlier) is blocked, there are several possible reactions, including hydrolysis of the ester, trans-esterification of the ester with another ester, and oxidation of the ester.

The hydrolysis of esters is a well-known reaction that involves as the initial step the attack of either a water molecule of a hydroxide ion at the carbonyl carbon. The use of highly hindered esters reduces the rate of acid- and base-catalyzed hydrolysis by blocking easy access to the carbonyl carbon of the ester. The rate of hydrolysis can be reduced if water is excluded from the oil; however, it is soluble to a concentration of approximately 0.5% in these lubricants. Hydrolysis typically results in the presence of acids (increased acid number) and partial esters, where the polyol has one or more hydroxyl groups present.

The process of trans-esterification can be best seen as changes in the acids attached to a given molecule. This process has long been known because, if a lubricant is prepared containing two initial esters with the same alcohol and different acids [i.e. $PE(nC_5)_4$ and $PE(nC_7)_4$], after a period of time, the acids are scrambled and $PE(nC_5)_3(nC_7)$, etc., are found in the mixture. This process does not normally result in a serious change in the properties of the lubricant and is actually thought to be beneficial by increasing the number of different components. However, if trans-esterification occurs with a phosphate ester from an additive, changes of reactivity are possible (**Figure 7**).

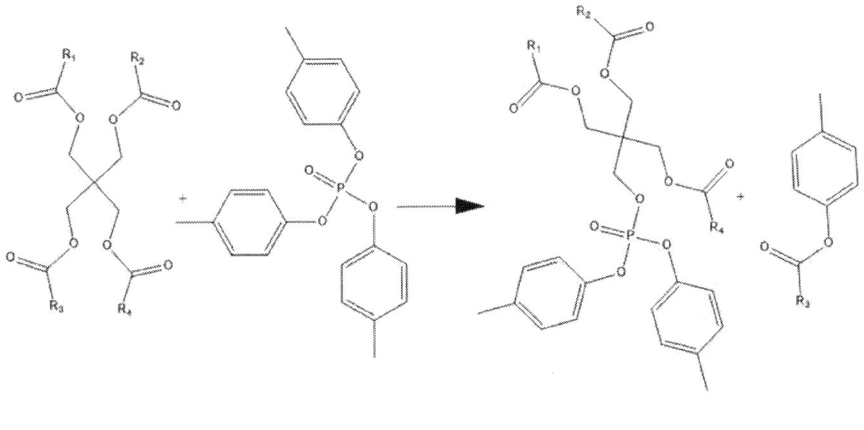

Figure 7. Reaction of a lubricant ester with phosphate esters by trans-esterification.

Possibly the most detrimental reaction to the properties of polyolester-based lubricants is high-temperature oxidation. The initial stages of the oxidation involve the attack of an alkyl peroxy radical on a methylene group of the ester. The position α to the carbonyl has been shown to be significantly more reactive than other methylene groups in the molecule [8]. After the initial attack, the reaction can progress to form anhydrides, which continue to react to form aldehydes, acids, and eventually high molecular weight compounds that can form a sludge in the engine (**Figure 8**) [9].

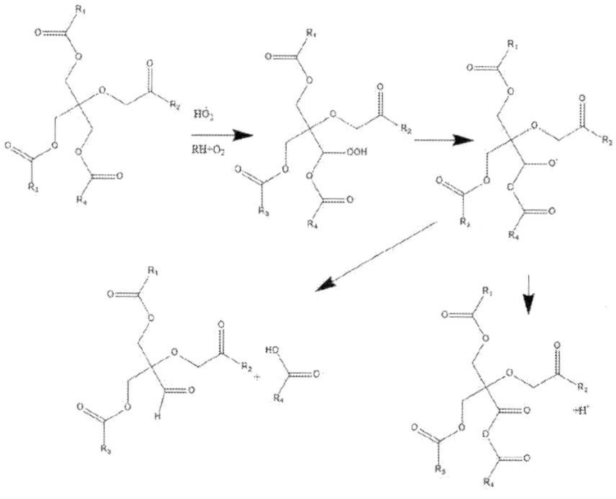

Figure 8. Thermal oxidation of PE ester at high temperature to give either anhydrides or an aldehyde and an acid.

The production of either an anhydride or an aldehyde is undesirable because they are reactive and have a tendency to polymerize, creating high molecular weight species. These reactions are eliminated in the presence of either BHT or an amine antioxidant until those concentrations are depleted.

4. Important lubricant properties

The development of a lubricant specification requires an examination of what properties are important and what methods will be used to demonstrate that the specification has been met. Many of the common properties, including the method used for the determination of the property, are shown below.

Viscosity (ASTM D 445)—Determines the thickness of the lubricant film that coats the bearing. Viscosity is measured at several temperatures and helps determine the maximum operating temperature of the engine. Changes in viscosity are observed as a lubricant degrades and these changes are a measure of lubricant health.

Pour Point (ASTM D 97)—The temperature at which a liquid becomes semi-solid or loses its flow characteristics. This property is related to the ability to service or start an engine in cold climates.

Flash Point (ASTM D 92)—The lowest temperature at which vapors of a fluid will ignite. Measuring a flash point requires an ignition source. At the flash point, the vapor may cease to burn when the ignition source is removed. This parameter is used as a way to control flammability, although understanding flammability requires a knowledge of many other parameters.

http://dx.doi.org/10.5772/62394

Total Acid Number (ASTM D 664)—A titration-based determination of the amount of acids contained in oil. Acids in the oil may cause corrosion of the bearings and valve guides. In ester-based lubricants, the acid number can indicate exposure of the lubricant to water, resulting in the hydrolysis of the oil. It can also indicate incomplete reaction when the oil was prepared. As the lubricant is used, the acid number of the used oil is also be an indicator of lubricant degradation.

Lubricant Compatibility (FED-STD-791, Method 3403)—A determination of hot aging miscibility of the candidate oil with other oils that might be encountered when the engine is in service. This test is necessary due to differences in the additive packages that various manufacturers use to achieve the required performance standard.

Elastomer Compatibility [Def Staf 05-50 (Part 61) Method 22]—A test used to determine the swell and deterioration of seal materials when soaked in oil for an extended time. Some high-stability basestocks and additive packages can cause the deterioration of certain seal materials over time.

Oxidation and Corrosion Stability (ASTM D 4636)—This test examines the bulk oil stability and the impact of breakdown products on certain representative metallic materials.

Thermal Stability and Corrosivity (FED-STD-791, Method 3411)—A quality control test for identifying contamination from non-aviation ester-based products.

Deposit Control—This test involves a full-scale heated bearing test rig. The test is run for 100 h, and the oil condition is evaluated and the cleanliness of the bearing is evaluated for the formation of deposits.

Load-Carrying Capacity or Boundary Film Lubricating Ability (FED-STD-791, Method 6509)—This property is measured with a Ryder Gear Test.

Foaming Tendency—Can be measured by either a static (ASTM D 892) or a dynamic method (Fed-STD-791, Method 3214). Oil can foam when subjected to the shear forces in the engine. Foaming is frequently due to impurities of various silicone compounds.

Sediment or Particulate Contamination (FED-STD-791, Method 310 or 3013)—This test measures the amount of suspended or dissolved solids in the oil. The low maximum ash content of MIL-PRF-23699 precludes the use of metal containing lubricant additives.

Hydrolytic Stability [Def Stan 05-50 (Part 61) Method 6]—Evaluates the storage stability and the stability of the oil in capped lubrication systems. Hydrolysis is a degradation reaction of the oil with water to regenerate the original alcohol and acids used to prepare the ester.

Evaporation Loss (AST D 972)—This test determines the amount of volatile components in the lubricant. A low evaporation loss means that too much oil will not be consumed during operation.

Shear Stability (ASTM D 2603)—Shear stability means that the lubricant will not lose viscosity due to the mechanical forces encountered in the operating engine.

Corrosion Inhibition (SAE ARP 4240)—Determined by ball corrosion testing, this attribute is only required for MIL-PRF-23699 oils of C/I class. Oils of the C/I class are primarily used in marine applications.

Storage Stability Tests—This test ensures that all of the lubricants remain miscible during extended storage periods of up to 3 years.

Acid Assay (FED-STD-791, Method 3500)—In this method, the component acids used in the preparation of the ester are measured. This is needed to ensure that different batches of the lubricant are similar to the original qualified lubricant.

Trace Metals—The trace metal content of new oils is controlled to provide a good baseline for the spectrographic oil analysis program.

Spectrographic Oil Analysis Program (SOAP)—The concentration of several wear metals (Fe, Ag, Cr, Al, Mg, Ti, Mo, and V) are measured regularly by atomic emission either using a rotating carbon electrode or by inductively coupled plasma (ICP-OES). As an engine is used, the concentrations of wear metals increase. If the normal wear pattern of the engine is known, changes in the metal concentrations that are not consistent with the pattern can be an indicator of impending problems. Excessive wear metals in the oil are an indication that some type of maintenance is needed and the metals present may provide an indication of the area of concern.

5. Lubricant specifications

Lubricant compatibility as a global problem requires standardization; however, the decision was made to not require a single supplier or a single highly specified composition. This would eliminate competition and would stop the development of better compositions. Instead, a series of detailed specifications were adopted for lubricants based on sets of properties. In the end, commercial jet aircraft lubricants are primarily governed by standard SAE-5780. Military jet aircraft generally fall under two standards based on different low temperature requirements. U.S. Air Force (USAF) aircraft typically use the MIL-PRF-7808 [10] specification because of the need for pumpability at lower temperatures. Naval aviation uses MIL-PRF-23699 [11], which has better high temperature properties. In most areas, MIL-PRF-23699 and SAE-5780 specifications are quite similar and many currently available products meet both of these specifications. These specifications undergo regular review and revision. In addition, MIL-PRF-7808L is now available in two grades with different viscosities: grade 3, a 3 cS viscosity fluid, and grade 4, a 4 cS viscosity fluid. The two grades require changes in the basestock composition, although similar chemistries are used. MIL-PRF-23699 is available in three types: standard, corrosion inhibiting (CI), and high thermal stability (HTS). The three MIL-PRF-23699 types differ primarily in the additive package used. A summary of the different lubricants is given in **Table 1**.

http://dx.doi.org/10.5772/62394

Specification	Grade or type	Viscosity (cS at 100°C)	Common use
MIL-PRF-7808L	Grade 3	3	Best low temperature properties
MIL-PRF-7808L	Grade 4	4	Improved high temperature properties but higher pour point
MIL-PRF-23699		5	Good viscosity index—not suitable for very cold climates
MIL-PRF-23699	CI	5	Includes CI—for use in maritime environments
MIL-PRF-23699	HTS	5	Best high temperature stability

Table 1. Common lubricant specifications used in aircraft engines.

Some of the specifications for each of the lubricants are included in **Table 2**.

Specification	MIL-PRF-7808L grade 3	MIL-PRF-7808L grade 4	MIL-PRF-23699	MIL-PRF-23699 CI	MIL-PRF-23699 HTS
Viscosity (cS)	3.0 min	1.1 min	4.90–5.40	4.90–5.40	4.90–5.40
205°C	11.5 min	4.0 min	23.0 min	23.0 min	23.0 min
100°C	1.7×10^4 max	17.0 min	13,000 max	13,000 max	13,000 max
40°C		2.0×10^4 max			
-40°C					
-51°C					
Flash point (°C)	210 min	210 min	246	246	246
Evaporative loss (%)	30 max	15 max	10 max	10 max	10 max
Foaming (static)	100 max	100 max	25 max	25 max	25 max
Volume (ml)	60 max	60 max	60 max	60 max	60 max
Collapse time (s)					
Deposition test	1.5 max	0.7 max	218°C	218°C	218°C
Rating	20 max	20 max	15 max	15 max	15 max
TAN change	100 max	100 max	120 max	120 max	60 max
(mg KOH/g)	100 max	100 max			
Viscosity at 40°C %					
Change Oil					
consumption (ml)					
Lead corrosion (g/m²)	9.3 max	9.3 max			
Corrosion test	**232°C**	**232°C**	±2.0	±2.0	±2.0
Silver (g/m²)	±4.5 max	±4.5 max			
Bronze (g/m²)	±4.5 max	±4.5 max			

Table 2. A summary of selected performance specifications for MIL-PRF-7808L [10] and MIL-PRF-23699 [11] lubricants.

6. Lubricant additives

Additives are a part of all lubricant systems because they can impart properties to the overall lubricant that the basestock does not possess. They can also allow longer lubricant lifetimes by eliminating certain modes of basestock decomposition. For example, when a lubricant is used, it is exposed to temperatures much higher than the bulk oil temperature for short periods of time, in the presence of oxygen. Additives can reduce the decomposition of the basestock by scavenging free radicals formed in the initial stages of the reaction. Aircraft lubricants typically contain the following groups of additives:

- Boundary lubrication additives (anti-wear, lubricity, and extreme pressure)
- Antioxidants
- Anti-corrosion additives
- Anti-foaming additives
- Metal deactivators
- Viscosity index improvers

The choices of additives are frequently limited based on the thermal stability of the additive, needs for long lifetimes, and tendencies of the additives to form deposits [12]. The mechanisms of action associated with each of the types of additive are described in the following sections.

6.1. Antioxidants

One of the primary decomposition reactions of lubricants is oxidation primarily by atmospheric oxygen. In lubricants derived from crude oil, there are normally enough naturally occurring sulfur compounds to inhibit oxidation for a period of time. Synthetic lubricants and some highly refined natural lubricants, however, must have additives to reduce or eliminate oxidation in the presence of oxygen. Oxidation of lubricants typically leads to an increase in viscosity and the formation of sludge and is the primary limit to the maximum bulk oil temperature.

Most classes of natural and synthetic lubricants are oxidatively degraded via a radical chain mechanism. In this mechanism, molecular oxygen attracts a hydrogen atom creating a free radical. The radical formed reacts with other lubricant molecules attracting hydrogen atoms or other groups, generally increasing the size of the molecule. The chain continues until it encounters another radical, resulting in chain termination.

Antioxidant additives can act in two different ways. First, they can react with oxygen to form a stable species. Second, the antioxidant can react with radicals formed, acting as a chain termination step. Among the most common antioxidants are hindered phenols and aromatic amines. The structure of these compounds is shown in **Figure 9**.

http://dx.doi.org/10.5772/62394

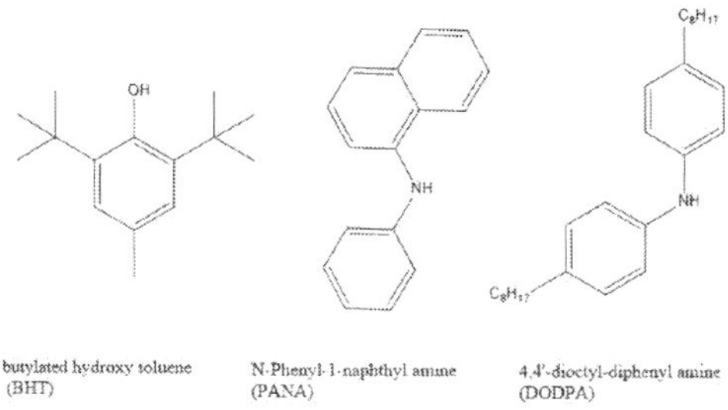

Figure 9. Structures of some common antioxidants.

As higher-temperature lubricants are being developed, these antioxidants are not sufficiently stable, which has led to the introduction of some high molecular weight polymeric antioxidants. Although the structures of these systems are proprietary, they solve similar problems and may involve similar chemistry.

All of the antioxidants above react to quench radical chain reactions. Because the amine antioxidants are common in aircraft lubricants, their mechanism will be discussed in more detail. It is commonly thought that the hindered amines are rapidly oxidized to the aminoxy radical. This radical reacts with alkyl radicals, trapping them to form N-alkoxy amines, which react with peroxyl radicals to give products and regenerate the aminoxy radical [13]. The oxidation of BHT has also been thoroughly studied. The initial oxidation by a radical removes a hydrogen atom forming a phenoxy radical that then reacts to form a quinoid, which is relatively stable.

6.2. Anti-corrosion additives

A corrosion inhibitor is a compound that decreases the rate of corrosion of a material, typically a metal or metal alloy. Corrosion inhibitors are of greater importance in the MIL-PRF-23699 CI lubricants primarily because of their intended use in more corrosive environments. Typically, with iron-based alloys (steel), corrosion occurs due to the presence of oxygen in contact with the steel, which results in the formation of rust. Corrosion inhibitors act by forming a passive layer on the surface of the alloy that protects the surface from further oxidation. Metal surfaces are typically covered with a layer of oxides and hydroxides formed when the metal is placed in contact with air. In some metals, such as aluminum, the oxide layer protects the surface from further oxidation (passivation). In iron-based alloys, however, the oxide layer is porous and further protection is needed.

Typical rust inhibitors include metal sulfonates and metal carboxylates. These additives act because they have an ionic head group that can bind to the metal oxide surface and a non-polar tail that forms a protective coating over the metal. These additives, however, are not ash less and may not qualify for the MIL-PRF-23699 CI standard. A product that avoids this problem consists of a mixture of an amine, carboxylic acid esters, and a phosphate ester. The mixture itself is proprietary; however, the MDS does at least give an approximate composition.

6.3. Anti-foaming additives

Foam is formed when air is trapped in a liquid forming bubbles. It is frequently observed when a stream of liquid re-enters the bulk liquid, such as when fuel is added to a fuel tank. Foams can result in reduced capacity and reduced pumping efficiency and prevent lubricants from effectively flowing. In aircraft applications, lubricants should form minimal amounts of foam and the foam formed should rapidly collapse. Typically, lubricant foaming is caused by contamination with high molecular weight silicone greases, which form foams that do not easily collapse. Foams collapse when the air bubbles merge forming larger bubbles that rise and the bubbles pop releasing the air at the surface [14].

Anti-foaming additives are all surface active agents that are insoluble in the lubricant. They typically are of low viscosity, which allows them to spread over the surface of the oil. Many also contain some solid particles dispersed in the additive. When a foam forms, the bubbles have structures similar to lipid bilayers called lamella. The anti-foaming additive locates itself at the air-liquid interface, maximizing the surface area of the foam bubbles. Smaller bubbles merge and the larger bubbles move toward the surface, causing the foam to collapse.

Aircraft lubricant foam control additives are typically moderate molecular weight silicone oils. It have been observed that silicone oils can act in many ways, causing foam in some cases and eliminating the foam in others depending on molecular weights and structures. The current anti-foaming additives have been shown to successfully eliminate foaming in the lubrication system.

6.4. Boundary lubrication additives

In aircraft applications, a lubricant must be effective in different lubrication regimes, including the fluid film lubrication regimes (hydrodynamic lubrication and elasto-hydrodynamic lubrication) and boundary lubrication. Hydrodynamic lubrication, where the two bearing surfaces are separated by a complete film of the lubricant, is generally desirable. As the shear stress on the bearing increases, the lubricant properties are not adequate and the film is no longer of sufficient thickness. Elasto-hydrodynamic lubrication occurs between rolling bodies [15]. These systems operate under stresses where a fluid film would not be expected to be maintained. The high pressure, however, causes a temporary increase in the viscosity of the lubricant in the contact area and a fluid film is maintained. Boundary lubrication involves direct contact between the surfaces of the bearings. Boundary lubrication is generally undesirable but is also unavoidable especially during start-up and shutdown. Boundary lubrication additives are required to reduce friction between and wear of the engine components until lubricant flow is adequate to maintain fluid film lubrication.

A number of boundary lubrication additives have been developed for various purposes and the vast majority are based on phosphorus or sulfur compounds. For aircraft applications, sulfur compounds are generally less desirable because of the lack of stability and the formation of strong acids on oxidation. Phosphorus compounds can include phosphate esters, thiophosphate esters, and metal thiophosphates. Metal thiophosphates and thiophosphate esters contain sulfur, and the metal dithiophosphates are more likely to form metal containing ash on decomposition. For aircraft applications, phosphate esters and, in particular, triaryl phosphates are the additives of choice.

Triaryl phosphates are characterized by three aromatic rings attached to the phosphorus through oxygen linkages. The most common commercial additives are tricresyl phosphate and butylated triphenyl phosphate. The structures of these compounds are shown in **Figure 10**.

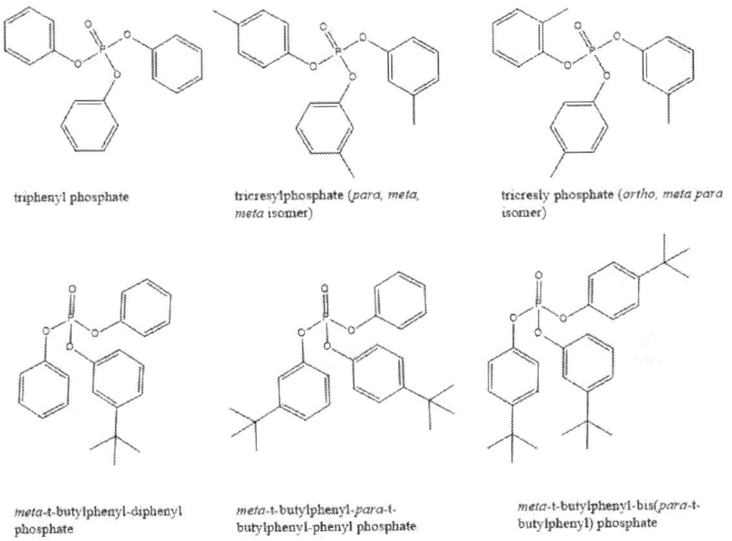

Figure 10. Structures of some common tri-aryl phosphate esters.

Phosphate esters are known to react at the surface of metals to form a coating that is both durable and lubricious. Phosphate esters have been shown to react with oxides or hydroxides on the metal surface to form a multilayer coating that is chemically bound to the surface. As the coating wears away, iron diffuses to the surface and more of the coating forms. The mechanism for coating formation begins with the oxide surface of the iron. A surface oxide displaces one of the alkyl groups from the phosphate ester giving a bound phosphate di-ester. Further reaction occurs to form iron polyphosphate, which has, at the surface, bound aryl groups (**Figure 11**) [16].

If a metal bearing is operated under conditions where little oxygen is present, there still appears to be a reaction to give a similar film. The principle difference in the film formation mechanism is to break a C-O bond to the aromatic ring, leaving a bound metal phosphate. The product of this reaction has been shown to add to another aromatic ring forming higher molecular weight phosphate esters. Under these circumstances, a coating of similar structure is still formed [17].

Advanced bearings under development have a very different surface chemistry. These bearings will be made from carburized or nitrided stainless steels. After machining of the bearing, it is heat treated in the presence of a carbon or nitrogen source. The resulting surfaces have little oxide or hydroxide present but a surface dominated by carbides or nitrides. It is unknown if phosphate ester additives will form a lubricious film on these surfaces. The non-heat-treated steels have been shown to react; however, heat-treated steals are currently under study [18]. The interaction of antioxidants, esters, and phosphate esters in the presence of various carbides has been shown to increase reactivity [19].

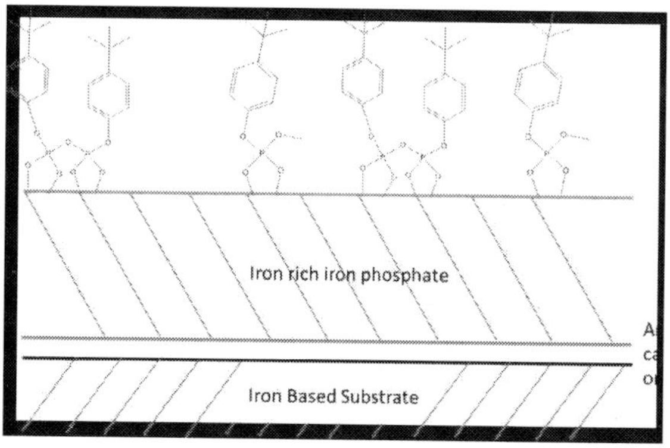

Figure 11. Structure of surface films formed by the reaction of phosphate esters with a steel bearing surface.

6.5. Metal atom deactivators

Metal ion deactivators are incorporated into lubricants to react with metal ions typically formed from the action of naturally occurring acids on the metallic parts of the lubrication system. The metal ions are of concern because of their catalytic effects, principally the tendency of copper ions to form a gummy copper mercaptide gel.

In aviation lubricants, the primary metal ion deactivator is benzotriazole. It is thought that benzotriazole reacts with metal atoms at the surface and in solution. On the surface, benzotriazole forms a passive layer that prevents further reaction at the metal surface. Dissolved metal ions react with the benzotriazole to form a complex, which reduces the reactivity of the soluble metal ion. An added benefit of benzotriazole as an additive is that it also acts as an antioxidant [20].

7. Future developments

In the quest for higher-performance turbine engines, a number of changes are under development. Possibly the most significant is the development of harder and more corrosion-resistant metal alloys such as Pyrowear and CSS22, among others, hard coatings such as TiC, and low-density silicone nitride–bearing materials. These materials are designed for high load carrying and higher temperature, which leads to challenges in lubricant stability. Harder and more stable coatings will allow for higher temperatures and loads adding stress to the lubricant system. It may be necessary to develop new basestocks that have wider liquid ranges and greater thermal stability.

More importantly, all of these new bearing materials have radically different surface chemistry than traditional steel-based bearings. The additive packages, particularly boundary lubrication additives, are dependent on the presence of an oxide or hydroxide surface coating to react with. The new bearing materials have a surface chemistry dominated by the presence of carbides or nitrides. The data are uncertain as to the level of surface coverage and the reactivity of the phosphate esters with these new surfaces.

Author details

David W. Johnson*

Address all correspondence to: djohnson1@udayton.edu

Department of Chemistry, University of Dayton, Dayton, Ohio, USA

References

[1] Poitz HA, Yungk RE. Aviation industry. In: Totten GE, editor. Handbook of Lubrication and Tribology: Volume I. Application and Maintenance. 2nd ed. CRC Press; 2006, pp. 26-1 to 26-15.

[2] Lansdown AR. Aviation lubricants. In: Mortier RM, Orszuliz ST, editors. Chemistry and Technology of Lubricants. Springer; 1994. pp. 223–236. DOI: 10.1007/978-1-4615-3554-6_9.

[3] David VW, Hughes JR, Reece D. Some lubrication problems of aviation gas turbines. Journal of the Institute of Petroleul. 1956;42:330.

[4] Hobson R. Synthetic aviation lubricants. Industrial Lubrication and Tribology. 1971; 23:7-11. DOI:10.1108/eb052958.

[5] Klamann D. Lubricants and Related Products: Synthesis, Properties, Applications, International Standards. Verlag Chemie; 1984. ISBN-13: 9780895731777.

[6] Korlipara VP, Bhamidipati VSKR, Potula SB, Rachapudi BNP, Singh AK. Polyol esters of medium chain fatty acids and process for preparation thereof. US Patent number: US8658813 B2. Published 2014. Assignee: Council of Scientific and Industrial Research.

[7] Gryglewicz S, Piechocki W, Gryglewicz G. Preparation of polyol esters based on vegetable and animal fats. Bioresource Technology. 2003;87:35–39. DOI: 10.1016/S0960-8524(02)00203-1.

[8] Niedzielski EL. Neopentyl polyol ester lubricants-bulk property optimization. Industrial & Engineering Chemistry Product Research and Development. 1976;15:53–58. DOI: 10.1021/i360057a010.

[9] Bakunin VN. A mechanism of thermooxidative degradation of polyol ester lubricants. Journal of Synthetic Lubrication. 1992;9:127–143. DOI:10.1002/jsl.3000090204.

[10] Beane GA, Gschwender LJ, Snyder CE, Shimski JT. Military aircraft propulsion lubricants—current and future trends. Journal of Synthetic Lubrication. 1986;3:263–291. DOI: 10.1002/jsl.3000030403.

[11] MIL-PRF-7808L, performance specification; lubricating oil, aircraft turbine engine, synthetic base (02 May 1997) [Superseding MIL-L-7808K]. 1997. Available at: http://everyspec.com/MIL-PRF/MIL-PRF-000100-09999/MIL-PRF-7808L_5699/ [Accessed 2016-01-31].

[12] MIL-PRF-23699G, performance specification: lubricating oil, aircraft turbine engine, synthetic base, NATO code numbers: O-152, O-154, O-156, and O-167 (13 Mar 2014) 23699 specs. 2014. Available at: http://everyspec.com/MIL-PRF/MIL-PRF-010000-29999/MIL-PRF-23699G_50301/ [Accessed 2016-01-31].

[13] Snyder CE, Gschwender LJ, Sharma SK. Long-term additive trends in aerospace applications. In: Rudnick LR, editor. Lubricant Additives: Chemistry and Applications. 2nd ed. CRC Press; 2009. pp. 637–646.

[14] Schwetlick K, Habicher WD. Antioxidant action mechanisms of hindered amine stabilisers. Polymer Degradation and Stability. 2002;78:35–40. DOI: 10.1016/S0141-3910(02)00116-7.

[15] Denkov ND. Mechanism of foam destruction by oil-based antifoams. Langmuir. 2004;20:9463–9505. DOI: 10.1021/la049676o.

[16] Hamrock BJ, Schmid SR, Jacobson O. Fundamentals of Fluid Film Lubrication. 2nd. ed. CRC Press; 2004. ISBN 9780824753719.

[17] Johnson DW, Hils JE. Phosphate esters, thiophosphate esters and metal thiophosphates as lubricant additives. Lubricants. 2013;1:132–148. DOI: 10.3390/lubricants1040132.

[18] Johnson DW, Morrow S, Forster NH, Saba CS. Vapor-phase lubrication: reaction of phosphate ester vapors with iron and steel. Chemistry of Materials. 2002;14:3767–3775. DOI: 10.1021/cm010921o.

[19] Johnson DW, Bachus M, Hils JE. Interaction between lubricants containing phosphate ester additives and stainless steels. Lubricants. 2013;1:48–60. DOI:10.3390/lubricants1020048.

[20] Johnson DW, Iaculla C, Hils JE. Reaction between polyol-esters and phosphate esters in the presence of metal carbides. Friction and Wear Research (FWR). 2013;1:1–9.

[21] Zhao G, Zhao Q, Wang X, Liu W. A novel benzotriazole derivative for enhancing the anti-oxidation properties of synthetic ester-based oils. Industrial Lubrication and Technology. 2014;66:353–359. DOI: 10.1108/ilt-12-2011-0112.

Section 2

Actuators and Controllers

Chapter 3

Active Flow Control by Using Plasma Actuators

Gabriele Neretti

Additional information is available at the end of the chapter

http://dx.doi.org/10.5772/62720

Abstract

Active flow control has recently received an increasing attention since it allows to directly manipulate the flow-field around a surface only when it is effectively requested. Aerodynamic plasma actuators supplied by a dielectric barrier discharge (DBD) can be used for this purpose. Usually, sinusoidal voltages in the range 5–50 kV peak and frequencies between 1 and 100 kHz are utilized to ignite this plasma typology. The surface discharge produced by these devices is able to tangentially accelerate the flow field by means of the electrohydrodynamic (EHD) interaction. DBDs generate non-thermal plasmas characterized by low input energies and limited temperature increments. Plasma actuators can be easily designed by following the shape of the aerodynamic body and can be used over heat-sensitive surfaces. These aerodynamic devices have demonstrated to produce boundary layer modifications with induced speeds up to 10 m/s. Their use over airfoils, flaps, and blades have shown the possibility to delay the transition between laminar to turbulent regime, to prevent flow separation enhancing lift and reducing drag. Moreover, the adoption of these actuators over landing gears and trailing edges may induce a noise reduction effect. Dielectric materials, electrodes configuration, and supplying waveforms are most relevant parameters to be considered to enhance actuator performance. On a parallel plane, on/off actuation strategy is a key point in the use of these devices when utilized over aerodynamic surfaces impinged within an external flow.

Keywords: Active flow control, Dielectric barrier discharge, Electrohydrodynamic interaction, Plasma physics, Noise mitigation

1. Introduction

The interest in flow control in the fluid dynamic domain has showed a quick growth in last 15 years [1] due to rapid development of sensors and actuators technologies. Even though passive control techniques are still attractive since they do not require an energy input, active control strategies have recently received more attention since they can be used in a selective way and can be operated only when it is effectively requested. Among different active techniques, plasma aerodynamic actuators are attractive because they present high dynamic responses due to the absence of moving parts, are characterized by low weight, are easy to build, are backward compatible with existing aerodynamic surfaces, and generate negligible aerodynamics interferences when they are switched off. When actuated, they can significantly modify the status of the boundary layer developing on the body surfaces. For this reason, they have been extensively studied for aeronautical applications to prevent flow separation enhancing lift and reducing drag [2, 3]. More recently, they have been used also to control friction drag by delaying transition [4, 5] or by oscillating the flow in spanwise direction [6], and to control global instabilities of the flow [2, 3, 7]. Due to these characteristics, the potential of plasma actuators has been extended to many other applications like for instance tip clearance flow control of turbines [8, 9], and wind turbine blades and holder [10, 11].

This brief review about DBD aerodynamic actuators is all but exhaustive. The main goals of this work are to give to the reader some basic knowledge about the physics of plasma actuators and to show most important applications of these devices into the active flow control domain.

2. DBD aerodynamic actuators: basic principles

Plasma aerodynamic actuators are based on the electrohydrodynamic (EHD) interaction generated by the so-called ionic wind. High electric fields can locally ionize the air. The produced heavy-charged species are accelerated by the applied electric field and, by means of collisions, they can yield momentum to the surrounding neutral gas. The force *f* per unit volume within the discharge can be yield by the following expression:

$$\mathbf{f} = \left(\mathbf{n}_{\mathrm{i}} - \mathbf{n}_{\mathrm{e}}\right)\overline{\mathbf{E}} \quad (1)$$

where E is the electric filed and n_i and n_e are ion and electron number density respectively.

First studies were conducted by using direct current (DC) corona discharges [2]. Main disadvantages of these actuators are low flexibility in electrodes configuration and in the supply system, and the transition into spark regime producing irreversible damage of the actuator itself.

Nowadays, the largest part of plasma aerodynamic actuators is based on the surface dielectric barrier discharge (SDBD). This typology of discharge allows to use several electrodes geometry, different supply voltage waveforms, and it prevents the transition into the arc regime due to the presence of the dielectric material. DBD actuators generate non-thermal plasmas [12], limiting power consumption and allowing their use over heat sensitive surfaces.

A classical DBD aerodynamic plasma actuator is constituted by a dissymmetric electrode pair separated by a dielectric slab (**Figure 1**). When high voltages (typically 10–50 kV at 1–100 kHz) are applied to the electrodes, a surface discharge is produced (**Figure 2**). Discharge appears to be macroscopically homogeneous to the unaided eye, but it is constituted by a sequence of micro-discharges [13] lasting typically for tens of nanoseconds with a repetition rate of several hundreds of megahertz. Fast ignition and quenching of the discharge can be inferred by voltage-current time behavior reported in **Figure 3**.

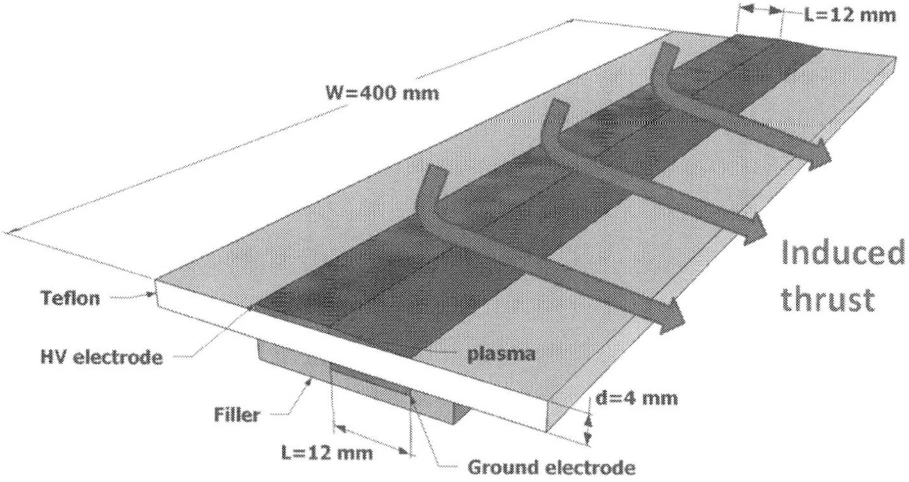

Figure 1. Schematic of a SDBD actuator for flow control.

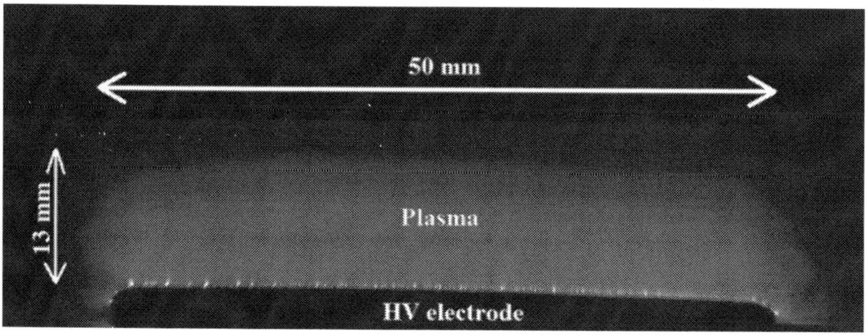

Figure 2. Top view image of SDBD discharge [14].

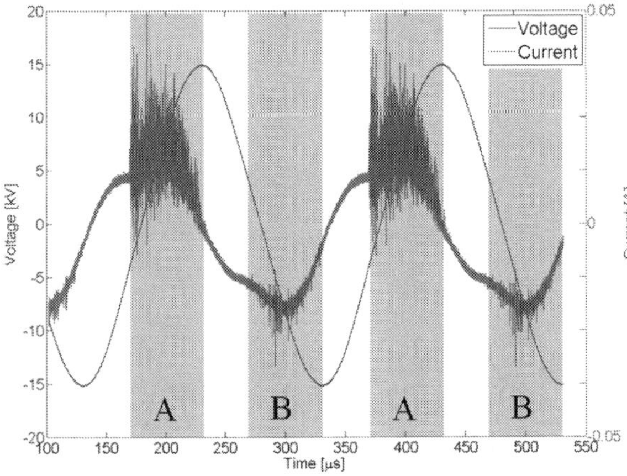

Figure 3. Voltage-current time behavior of a SDBD [15].

The presence of the plasma and the particular electrode configuration induce a jet tangential to the actuator wall, similar to a classic blowing technique [13]. These jets can modify the aerodynamic boundary layer, increasing flow momentum, at least in the near-wall region above the surface. A large number of experimental [14–25] and numerical [26–47] works have been done in last decade to understand basic physical phenomena involved in the EHD interaction.

A typical velocity profile induced by the tangential wall jet is shown in **Figure 4**. Measurements have been carried out by a glass Pitot tube positioned 2 mm downstream with respect plasma extension and moved parallel to the actuator surface. Maximum velocity of about 5–6 m/s are usually reached.

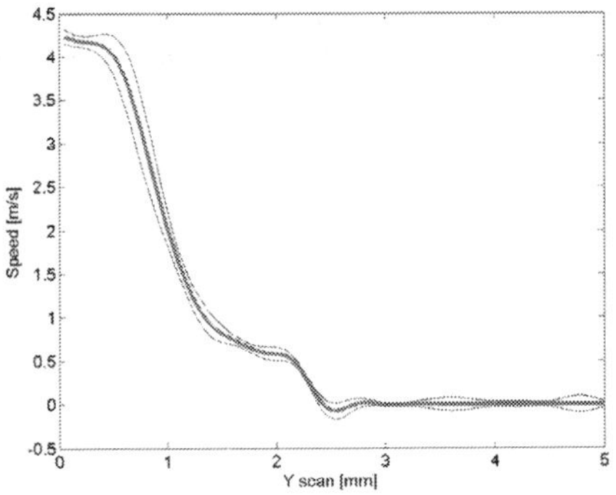

Figure 4. Typical pitot velocity profile of a DBD plasma actuator (red line). Blue lines represent standard deviation [48].

Fast dynamics of these actuators is underlined in **Figure 5** where Schlieren images of the induced jet developing during ignition of the discharge are shown. Steady state (d) is typically reached after 400 ms after discharge ignition.

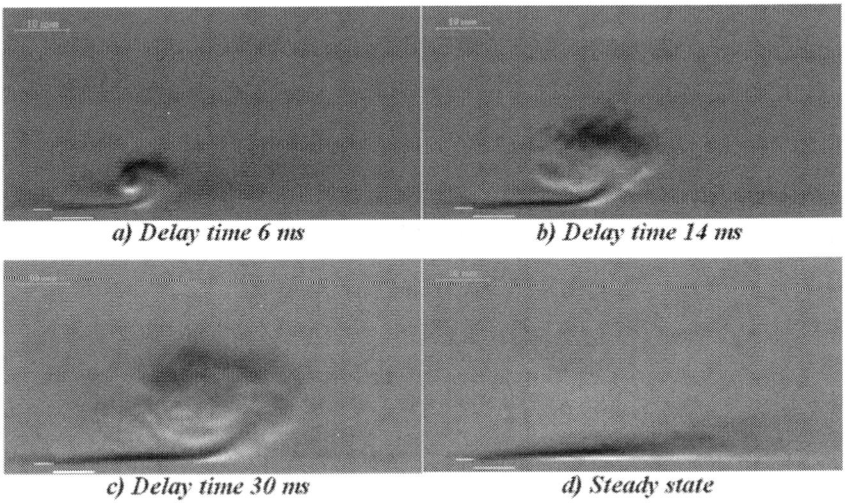

Figure 5. Schlieren images of induced wall jet during ignition phase of the discharge (a, b, and c) and steady state operation (d) [48].

As already introduced [26–47], EHD interaction produced by DBD plasma actuators has been numerically investigated a well. Main issues are related to the simulation of the discharge and its interaction with surrounding gas. A 'realistic' simulation of a DBD streamer should take into account gas ionization, plasma chemistry, thermal fluxes, diffusion of neutral and charged species, and charge deposition over dielectric surfaces. Time steps of about tens of picoseconds and mesh size in the order of few micrometers are usually required to reproduce the formation of a plasma filament lasting for tens of nanoseconds and featured with characteristic length of about few millimeters. Several coefficients related to diffusion, secondary electrons emission, reaction rates, and electrons attachment are often not easy to estimate or to recover from existing literature. The numerical model leads to the temporal/spatial evolution of the body force produced by the discharge. The second step is the coupling between the discharge and the surrounding gas. This is another numerical challenge because characteristic times and lengths of the discharge are several orders of magnitude smaller with respect those related to the fluid-dynamic domain. Several authors [26–36] followed this computational strategy obtaining interesting results in good agreement with experimental fluid-dynamic effects. In **Figure 6a**, integrated horizontal body force as a function of the applied voltage is shown [36]. The two half periods produce different force dynamics and magnitude. In **Figure 6b**, the time-averaged horizontal force distribution above the actuator surface is depicted [33]. Highest values have been obtained close to plasma end.

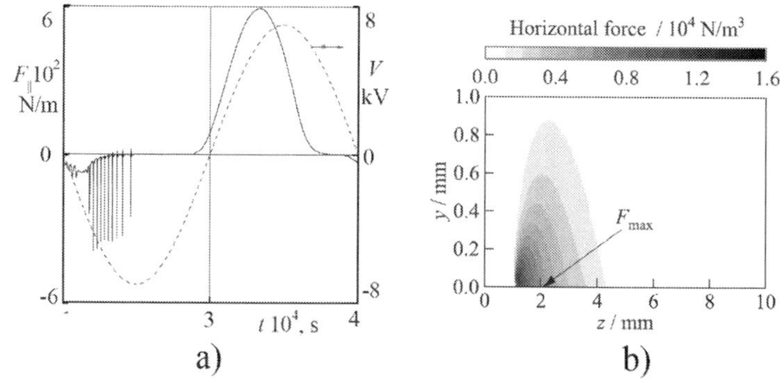

Figure 6. Time evolution of the EHD body force (a) [36], and spatial distribution of the average horizontal force (b) [33].

A simpler approach is to estimate the electron number density [37–42], on the basis of experimental measurements or theoretical calculations, and to evaluate EHD body force using Equation 1. This force is subsequently utilized as input parameter in the Navier-Stokes equation solver. A spatial/temporal average value of the body force can also be obtained by means of experimental results [43–47]. This method allows to use standard fluid-dynamic solvers avoiding difficulties arising from the couple between plasma and surrounding gas.

Both numerical and experimental works demonstrated the capability of plasma actuators to produce a thrust able to move surrounding gas in a desired direction. In order to increase both induced speed and region in which the actuator manifests its influence, a series of DBD actuators can be manufactured (**Figure 7**). Each actuator of this multi-electrode arrangement contributes to increase the induced speed.

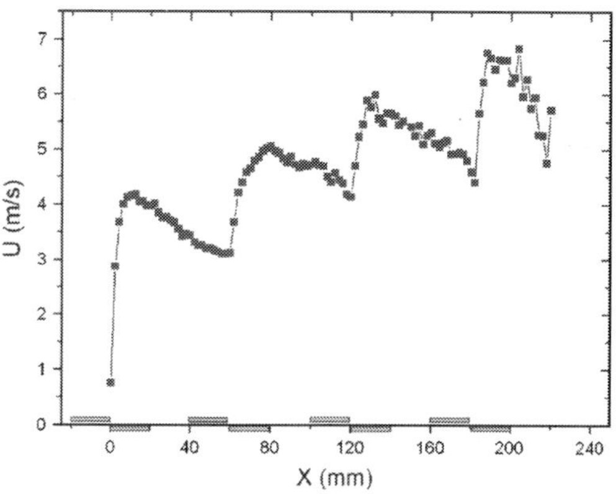

Figure 7. Horizontal velocity profile (y = 0.5 mm) above four DBD actuators in series [13].

On a parallel plane, voltage waveform shape (arbitrary or nanopulsed signals) can enhance thrust of the induced jet, increasing effectiveness of actuators [15, 49–56]. The use of arbitrary waveforms can enhance plasma ignition phases in which induced thrust is higher. In this way, induced speed (**Figure 8**) and efficiency (**Figure 9**) can be both increased [15]. Nanopulsed discharges are able to generate energetic plasmas lasting for few nanoseconds. During the ignition of the discharge, a high amount of energy is deposited within the gas inducing local shock waves able to significantly modify the boundary layer of an incoming flow [56].

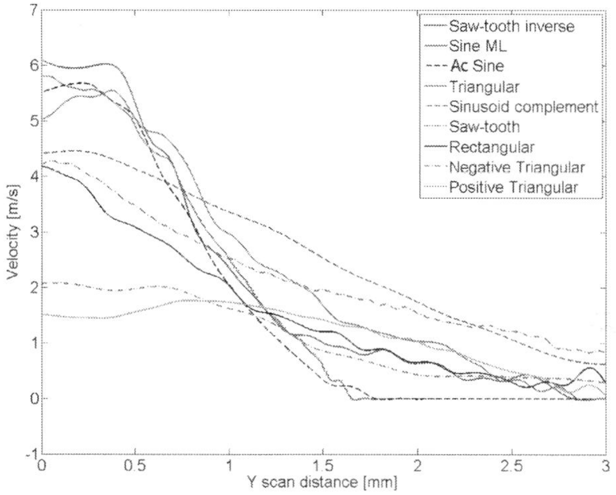

Figure 8. Pitot velocity profiles obtained close to plasma extension for different supplying voltage waveforms [15].

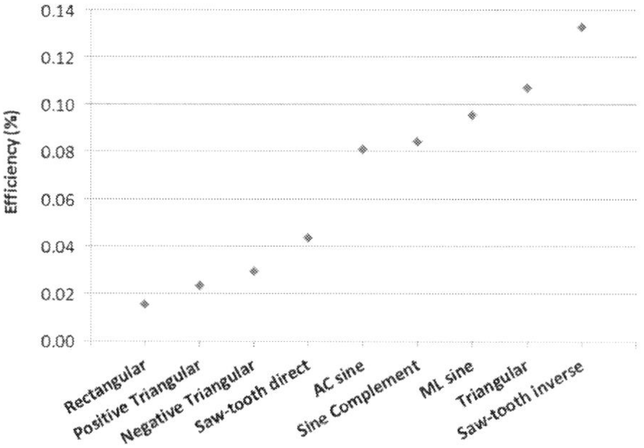

Figure 9. Electric into kinetic conversion efficiency for a DBD actuator driven with different voltage waveforms [15].

Another possibility to increase actuator performance is the adoption of a third exposed electrode supplied with high-voltage DC fields. In this way, a sliding DBD is generated. Depending on the sign of the DC field, it is possible to modify the morphology of the induced ionic wind [57, 58].

DBD actuators can be arranged to produce normal or vectorized jets, too. These devices can be used to mimic classical synthetic jets and are usually called plasma synthetic jet actuators (PSJA). Typical geometries are annular and linear one [48, 59–62]. In both cases, tangential wall jets collide merging in a unique-induced jet able to generate perturbations far away from the actuator surface (**Figures 10** and **11**). When the linear configuration is adopted, by supplying exposed electrodes with different voltages, it is possible to produce an induced jet with an arbitrary inclination. Three-dimensional flows can be also produced by using a classical DBD actuator, but with an exposed electrode characterized with serpentine or serrated geometries, instead of the usual linear one [63, 64]. This particular electrode geometry generates small tangential jets that collide and propagate downstream pushed by the EHD interaction body force.

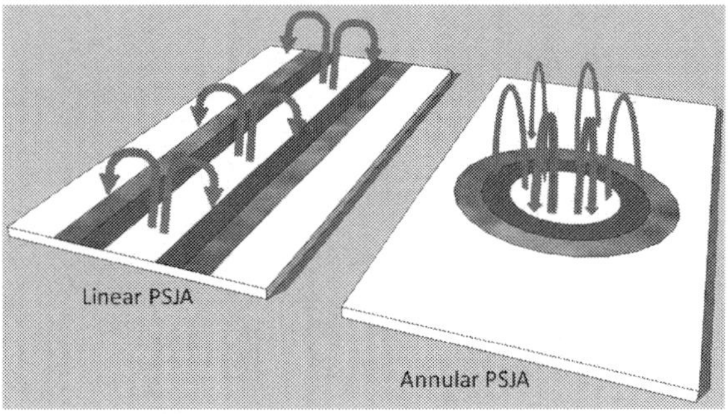

Figure 10. Induced normal jet of linear (left-hand side) and annular (right-hand side) PSJA.

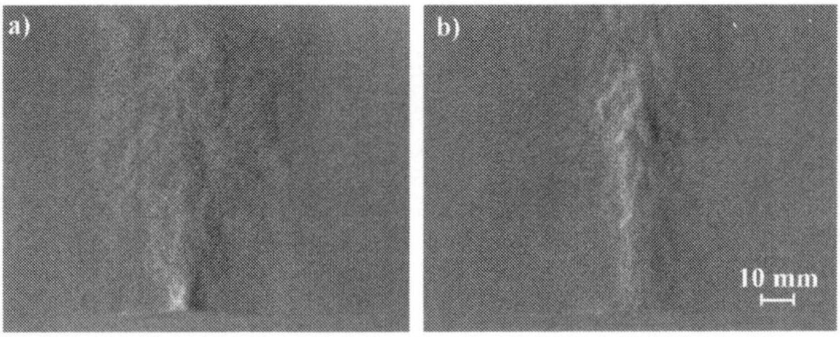

Figure 11. Schlieren images of PSJAs of linear (a) and annular (b) geometries. Electrode distance (a) and diameter (b) is 30 mm.

http://dx.doi.org/10.5772/62720

3. DBD aerodynamic actuators: flow manipulation

In the last decade, DBD aerodynamic actuators have been extensively studied in the active flow control domain. Position and actuation strategies of these devices are key points in their effectiveness in flow control over aerodynamic surfaces. Many studies have been accomplished over airfoils [65–72], diffusers [73, 74], and wind [10, 11 and 75] and turbine [8, 9] blades, in order to enhance fluid-dynamic performance. Moreover, the adoption of these devices over landing gears and trailing edge surfaces have demonstrated the possibility to obtain a noise reduction effect [76–78].

DBD actuators can be usually positioned over aerodynamic surfaces in spanwise and streamwise directions. In the former, the induced body force is in the same direction as the incoming flow. In the latter, induced thrust is perpendicular to the free stream direction. In this case, the composition of these two flows produces vorticities propagating in the downstream direction.

In **Figure 12**, a NACA 0015 airfoil equipped with four spanwise DBD plasma actuators is shown. The four plasma regions are clearly visible in the figure as bluish strips. In this configuration, actuators produce a tangential wall jet directed downstream. The effect of this plasma device in the recovery of stall condition is depicted in **Figure 13**, where smoke visualization is reported. Experiments show how the most effective actuator is the one positioned on the airfoil leading edge, just before the region where separation occurs [66, 67, 79, 80].

Figure 12. Airfoil equipped with four spanwise plasma actuators [65].

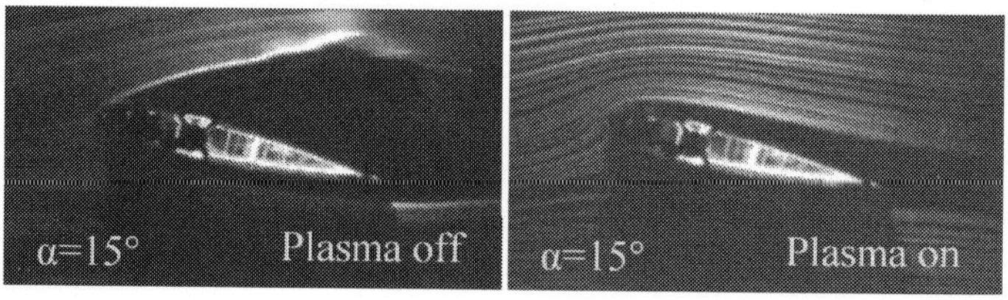

Figure 13. Flow structure around NACA0015 airfoil at Re = 15000 without (left-hand side) and with (right-hand side) plasma actuation [65].

When DBD actuators are streamwise mounted, induced and incoming flows combine together producing vorticities propagating in the downstream direction (**Figure 14**). Such a device can be used as a plasma vortex generator (PVG)

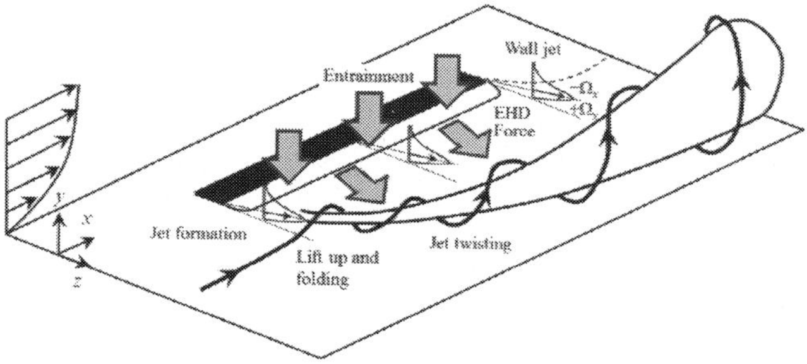

Figure 14. Vortex formation mechanism induced by a streamwise plasma actuator [8].

Changes in drag coefficient (C_D) and lift coefficient (C_L) by using a PVG are reported in **Figure 15**. At low Reynolds numbers, actuator effectiveness is very pronounced, leading to noticeable C_D reduction with a parallel increment in the stall angle.

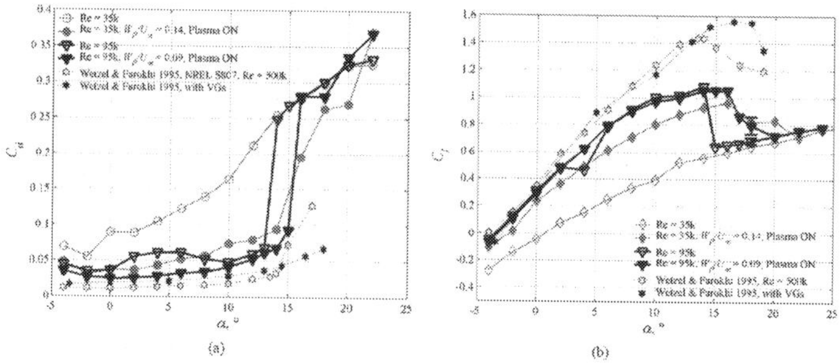

Figure 15. Drag (a) and lift coefficient enhancement (b) with counter-rotating PVGs on a NACA 4418 airfoil [8].

http://dx.doi.org/10.5772/62720

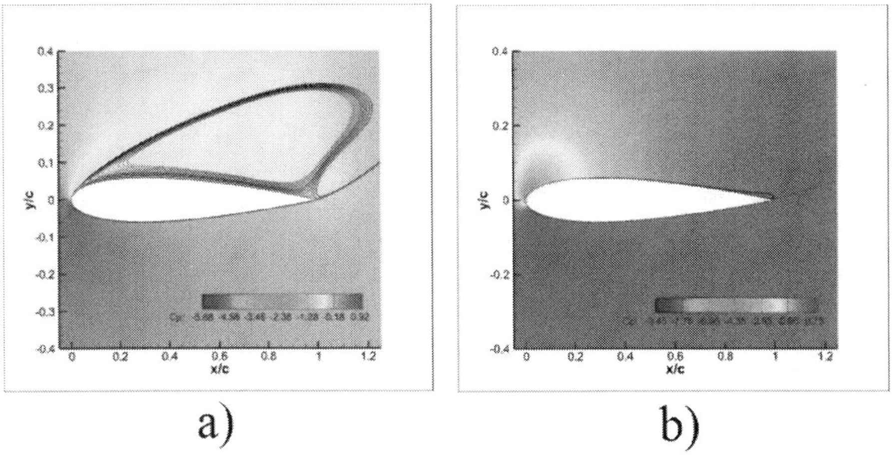

Figure 16. Pressure coefficient contours with streamlines over an airfoil with 15° angle of attack: plasma actuator off (a) and on (b) [47].

Effectiveness in flow control by means plasma actuators has been demonstrated by numerical works, too [47, 81–87]. In **Figure 16**, pressure coefficient contours with streamlines obtained over an airfoil with a 15° angle of attack are displayed. In **Figure 16a**, plasma actuator is switched off, and separation occurs. When the plasma device is activated, reattachment of the flow is achieved (**Figure 16b**).

Plasma actuators have demonstrated to strongly improve their ability in flow manipulation when operated with a duty cycle strategy [47, 66, 71, 72, 88]. With this approach, the DBD device is turned on and off with intermittence by following a particular duty cycle frequency. This frequency is usually chosen in the range 5–100 Hz and it is strictly related to natural vorticities developing over the aerodynamic surface. The percentage ratio between the period in which discharge is fed and the whole duty cycle period is called duty cycle percentage. If it

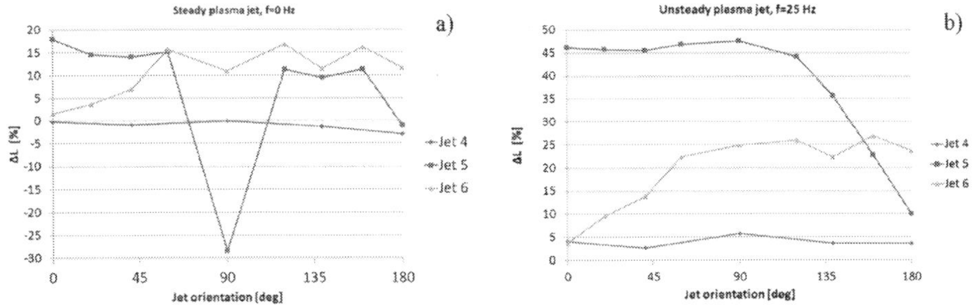

Figure 17. Lift recovery in percentage for different actuators location with a free stream velocity of 11 m/s: steady operation (a) and duty cycle operation (b) [66].

is fixed to 50%, it means that discharge is ignited half of the time. On a parallel plane, this actuation strategy leads to lower power consumption.

Figure 17 shows lift recovery in percentage of a stalled NACA0015 airfoil equipped with spanwise vectorized actuators located in different airfoil positions [66]. Jet 5 is generated by a DBD actuator located in the leading edge. When actuators are continuously operated, lift increments are limited to about 15%. When operated with a duty cycle strategy, lift increments are close to 50%.

Plasma actuators can also be used to reduce noise induced by aerodynamic surfaces, especially by landing gears and trailing edges [76–78]. Studies on bluff bodies have demonstrated the ability of DBD actuators to reduce downstream turbulence, leading to the suppression of particular tones or to an overall noise mitigation up to 4 dB.

4. Conclusion

Studies carried out in the last decades have demonstrated the possibility to use DBD aerodynamic actuators for active flow control purposes. In this last section, main advantages and disadvantages in the use of these plasma devices will be summarized.

Pros

- Possibility to locate these actuators in different positions on a surface and over existing aerodynamic bodies.
- Negligible aerodynamic interferences when they are not active.
- Fast actuation times.
- Low power consumption.
- Possibility to tune the on/off actuation strategy depending on the particular fluid dynamic condition.

Cons

- Induced velocity below 10 m/s.
- Low electric into kinetic conversion efficiency.
- Use of potentially hazardous high voltages and electromagnetic noise generation.
- Effectiveness in flow control for Reynolds numbers typically below few hundred thousand.

Disadvantages already introduced will be overcome only when a more detailed knowledge of basic interactions between discharge a neutral gas will be achieved. When higher induced speeds will be available, new and more effective flow control strategies will be developed.

Author details

Gabriele Neretti

Address all correspondence to: gabriele.neretti@unibo.it

Electrical, Electronic Engineering Department Guglielmo Marconi, University of Bologna, Bologna, Italy

References

[1] M. Gad-El-Hak. Flow Control. Cambrige: Cambrige University Press; 2000.

[2] E. Moreau. Airflow control by non-thermal plasma actuator. J. Phys. D: Appl Phys. 2007;40:605. DOI: 10.1088/0022-3727/40/3/S01

[3] J.J. Wang, K. Choi, L. Feng, T. N. Jukes, R. D. Whalley. Recent developments in DBD plasma flow control. Progress in Aerospace Sciences. 2013;62:52–78. DOI:10.1016/j.paerosci.2013.05.003

[4] S, Grundmann, C. Tropea. Experimental transition delay using glow-discharge plasma actuators. Exp. Fluids. 2007;42:653–657.

[5] S. Grundmann, C. Tropea. Active cancellation of artificially introduced Tollmien-Schlichting waves using plasma actuators. Exp. Fluids. 2008;44:795–806.

[6] S.P. Wilkinson. Investigation of an oscillating surface plasma for turbulent drag reduction. In: Proceedings of the 41st aerospace sciences meeting and exhibit; Reno, USA. AIAA; 2003. AIAA 2003–1023

[7] G. Touchard. Plasma actuators for aeronautical applications-state of art review. International Journal of Plasma Environmental Science and Technology. 2008;2(1):1–25.

[8] Choi, K.–S., Jukes, T.N., Whalley, R.D., Feng, L.H., Wang, J.J., Matsunuma, T. and Segawa, T.. Plasma Virtual Actuators for Flow Control. Journal of Flow Control, Measurement & Visualization. 2005;3:22–34. DOI:10.4236/jfcmv.2015.31003.

[9] Thomas C. Corke, Flint O. Thomas, and Junhui Huang. Documentation and Control of Flow Separation on a Low Pressure Turbine Linear Cascade of Pak-B Blades Using Plasma Actuators. In: NASA/CR-2007–214677, editor. University of Notre Dame, Aerospace and Mechanical Engineering Department, . 2007.

[10] Robert C. Nelson Thomas C. Corke, Hesham Othman, Mehul P. Patel, Srikanth Vasudevan, and Terry Ng. A Smart Wind Turbine Blade Using Distributed Plasma

Actuators for Improved Performance. In: 46th Aerospace Sciences Meeting; January 7–10; Reno, Nevada (USA). 2008. DOI: AIAA-2008–1312.

[11] D. Greenblatt, A. Ben-Harav, H. Mueller-Vahl. Dynamic Stall Control on a Vertical-Axis Wind Turbine Using Plasma Actuators. AIAA Journal. 2014;52:456–462.

[12] U Kogelschatz. Dielectric-barrier Discharges: Their History, Discharge Physics, and Industrial Applications. Plasma Chemistry and Plasma Processing. 2003; 23(1):1–46.

[13] N. Bénard and E. Moreau. Electrical and mechanical characteristics of surface AC dielectric barrier discharge plasma actuators applied to airflow control. J. Experiments in Fluids. 2014;55:1846:1–43. DOI: 10.1007/s00348–014-1846-x.

[14] A. Cristofolini, C.A. Borghi, and G. Neretti. Charge distribution on the surface of a dielectric barrier discharge actuator for the fluid-dynamic control. J. Appl. Physics. 2013;113(143307)DOI: 10.1063/1.4799159.

[15] C.A. Borghi, A. Cristofolini, G. Grandi, G. Neretti, and P. Seri. A plasma aerodynamic actuator supplied by a multilevel generator operating with different voltage waveforms. Plasma Sources Science and Technologies. 2015; 24(4). DOI: 10.1088/0963–0252/24/4/045018.

[16] C.A. Borghi, M.R. Carraro, A. Cristofolini, G. Neretti. Electrohydrodynamic interaction induced by a dielectric barrier discharge. J. Appl. Physics. 2008;103DOI: 10.1063/1.2888354

[17] A. Cristofolini, G. Neretti, F. Roveda, and C.A. Borghi. Schlieren imaging in a dielectric barrier discharge actuator for airflow control. J. Appl. Physics. 2012;111(033302)DOI: 10.1063/1.3682488J.

[18] Pons, E. Moreau, G. Touchard. Electrical and aerodynamic characteristics of atmospheric pressure barrier discharges in ambient air. In: Proc. ISNTPT; 2004. p. 370–410.

[19] A. Cristofolini, G. Neretti, C.A. Borghi. Effect of the charge surface distribution on the flow field induced by a dielectric barrier discharge actuator. J. Appl. Physics. 2013;114(073303)DOI: 10.1063/1.4817378

[20] G. Neretti, A. Cristofolini, C.A. Borghi, A. Gurioli, and R. Pertile. Experimental results in DBD plasma actuators for air flow control. IEEE Transactions on Plasma Science. 2012;40:1678–1687. DOI: 10.1109/TPS.2012.2191801

[21] F.O. Thomas, T.C. Corke, M. Iqbal, A. Kozlov, and D. Schatzman. Optimization of Dielectric Barrier Discharge Plasma Actuators for Active Aerodynamic Flow Control. AIAA Journal. 2009;47:2169–2178. DOI: 10.2514/1.41588

[22] T.C. Corke, C.L. Enloe, and S.P Wilkinson. Dielectric barrier discharge plasma actuators for flow control. Annual Review of Fluid Mechanics.2010;42: 505 - 529.

[23] G.I. Font, C.L. Enloe, and T.E. McLaughlin. Plasma volumetric effects on the force production of a plasma actuator. AIAA Journal. 2010;48(9):1869 - 1874.

[24] M. Forte, J. Jolibois, E. Moreau, G. Touchard, and M. Cazalens. Optimization of a dielectric barrier discharge actuator by stationary and non-stationary measurements of the induced flow velocity: application to airflow control. Experiments in Fluids.43(6): 917 - 928, 2007.

[25] M. Kotsonis. Diagnostics for characterisation of plasma actuators. Measurement Science and Technology. 2015; 26(9): 092001. DOI: 10.1088/0957–0233/26/9/092001.

[26] J.P. Boeuf, Y. Lagmich, and L.C. Pitchford. Contribution of positive and negative ions to the elecrtrohydrodynamic force in a dielectric barrier discharge operating in air. J. Appl. Phys. 2009;106 (023115)

[27] Hiroyuki Nishida, and Takashi Abe. Validation Study of Numerical Simulation of Discharge Plasma on DBD Plasma Actuator. In: 42nd AIAA Plasmadynamics and Lasers Conference; 27 – 30 June; Honolulu, Hawaii. 2011. AIAA 2011–3913.

[28] Meenakshi Mamunur, Uwe Kortshagen, Douglas Ernie, and Terrence Simon. Plasma actuator simulation: Force contours and dielectric charging characteristics. In: 48th AIAA Aerospace Sciences Meeting Including the New Horizons Forum and Aerospace Exposition; 4 – 7 January; Orlando, Florida. 2010. AIAA 2010–1221

[29] Likhanskii, A., Shneider, M., Macheret, S., Miles, R.. Modeling of dielectric barrier discharge plasma actuators driven by repetitive nanosecond pulses. Physics of Plasmas . 2007;14(073501)

[30] Likhanskii, A. Simulation Tool for Dielectric Barrier Discharge Plasma Actuators at Atmospheric and Sub-Atmospheric Pressures. NASA/CR-2012–217260. 2012.

[31] J.S. Shang, P.G. Huang. Surface plasma actuators modeling for flow control. Progress in Aerospace Sciences. 2014;67:29–50. DOI: 10.1016/j.paerosci.2014.01.001

[32] M. Abdollahzadeh, J.C. Pascoa, P.J. Oliveira. Implementation of the classical plasma-fluid model for simulation of dielectric barrier discharge (DBD) actuators in Open-FOAM. Computers & Fluids. 2016;108:77–99. DOI: 10.1016/j.compfluid.2016.01.012

[33] S. L. Chernyshev , A. P. Kuryachii, D. A. Rusyanov, V. V. . Numerical modelling of plasma multi-actuator system. Blucher Mechanical Engineering Proceedings. 2014;1(1): 3026–3037. DOI: 10.5151/meceng-wccm2012–19129.

[34] Likhanskii, M.N. Shneider, D.F. Opaits, R.B. Miles and S.O. Macheret. Numerical modelling of DBD plasma actuators and the induced air flow. In: 38th AIAA Plasma-dynamics and Lasers Conference; 25 - 28 June 2007, Miami, FL. AIAA 2007–4533.

[35] Hiroyuki Nishida, Taku Nonomura, and Takashi Abe. Characterization of Electrohy-drodynamic Force on Dielectric-Barrier-Discharge Plasma Actuatorusing Fluid Simulation. International Journal of Mechanical, Aerospace, Industrial, Mechatronic and Manufacturing Engineering. 2012;6(11): 2324–2328. DOI:scholar.waset.org/ 1999.8/9312.

[36] S. L. Chernyshev, A.P. Kuryachii, S.V. Manuilovich, D.A. Rusyanov, and V.V. Skvortsov. On the possibility of laminar flow control on a swept wing by means of plasma actuators. Progress in Flight Physics. 2015;7:169–190. DOI: 10.1051/eucass/201507169.

[37] M. Abdollahzadeh, J. Páscoa, P. Oliveira. Numerical Modeling of Boundary Layer Control Using Dielectric Barrier Discharge. In: MEFTE 2012 - IV Fluid Mechanics, Thermodynamics and Energy National Conference; 28-29 May 2012; Lisbon, Portugal. 2012.

[38] Y. Suzen, P. Huang, J.D. Jacob.Ashpis. Numerical simulations of plasma based flow control applications. In: 35th AIAA Fluid Dynamics Conference and Exhibit, Fluid Dynamics and Co-located Conferences; 6-9 June, 2005; Toronto, Canada. 2005. DOI: 10.2514/6.2005-4633

[39] Y. Suzen, G. Huang, D. Ashpis. Numerical simulations of flow separation control in low pressure turbines using plasma actuators. In: 45th AIAA Aerospace Sciences Meeting and Exhibit; 8 - 11 January 2007; Reno, Nevada, AIAA-937-2007.

[40] C. Enloe, T.E. McLaughlin, R.D. vanDyken, K. Kachner, E.J. Jumper, T. C Corke. Mechanisms and responses of a single dielectric barrier plasma actuator: plasma morphology. AIAA Journal . 2004; 42: 589-594.

[41] D.M. Orlov. Modelling and simulation of single dielectric barrier discharge plasma actuators. [PhD Thesis]. 2006 D.M.Orlov.

[42] B.E. Mertz, T.C. Corke, Time-dependent dielectric barrier discharge plasma actuator modeling. In: 47th AIAA Aerospace Sciences Meeting; Orlando, FL. AIAA Paper, vol. 1083, 2009.

[43] N. Benard, A. Debien, E. Moreau. Time-dependent volume force produced by a non-thermal plasma actuator from experimental velocity field. J. Phys. D: Appl. Phys; 46 (245201); 2013.

[44] A. Kurz, S. Grundmann, C. Tropea, M. Forte, A. Seraudie, O. Vermeersch, D. Arnal, N. Goldin, R. King. Boundary Layer Transition Control using DBD Plasma Actuators. Journal Aerospace Lab. 2013; AL06-02.

[45] A. Cristofolini, G. Neretti, C. Borghi. Experimental and numerical investigation on the surface charge distribution in a Dielectric Barrier Discharge fluid-dynamics plasma actuator. In: 43rd AIAA Plasmadynamics and Lasers Conference; 25-28 June, 2012; New Orleans, Louisian. 2012. p. AIAA-2012-3090.DOI: http://dx.doi.org/10.2514/6.2012-3090.

[46] A. Criscione, B. Frohnapfel, C. Tropea. Preliminary Numerical Assessment of Turbulent Flow Control with Plasma Actuators. In: A. Dillmann, G. Heller, H. P. Kreplin, W. Nitsche, I. Peltzer. editors. New Results in Numerical and Experimental Fluid Mechanics VIII. Springer Berlin Heidelberg; 2013. p. 201-208. DOI: 10.1007/978-3-642-35680-3.

[47] T. K. West IV. Numerical investigation of plasma actuator configurations for flow separation control at multiple angles of attack [Master thesis]. 2012. Paper 5197.

[48] G. Neretti, A. Cristofolini, and C.A. Borghi. Experimental investigation on a vectorized aerodynamic dielectric barrier discharge plasma actuator array. J. Appl. Physics. 2014; 115(163304). DOI: 10.1063/1.4873896.

[49] N. Balcon, N. Benard, Y. Lagmich, J.P. Boeuf, G. Touchard, E. Moreau. Positive and negative sawtooth signals applied to a DBD plasma actuator influence on the electric wind. Journal of Electrostatics. 2009;67:140–145.

[50] Benard N., E. Moreau. Role of the electric waveform supplying a dielectric barrier discharge plasma actuator. Applied Physics Letters. 2012; 100(9). ref. 193503.

[51] M. Kotsonis and S. Ghaemi. Performance improvement of plasma actuators using asymmetric high voltage waveforms. J. Phys. D: Appl. Phys. 2012; 45 (045204). DOI: 10.1088/0022-3727/45/4/045204.

[52] F.A. Dragonas, G. Neretti, P. Sanjeevikumar, and G. Grandi. High-Voltage High-Frequency Arbitrary Waveform Multilevel Generator for DBD Plasma Actuators. IEEE Transactions on Industry Applications. 2015; 51(4): 57-61. DOI:0.1109/TIA. 2015.2409262.

[53] D. Opaits, A. Likhanskii, G. Neretti, S. Zaidi, M. Shneider, R. Miles, and S. Macheret. Experimental Investigation on Dielectric Barrier Discharge Plasma Actuators Driven by Repetitive High-Voltage Nanosecond Pulses with DC or Low Frequency Sinusoidal bias. J. Appl. Physics. 2008; 104. DOI: 10.1063/1.2968251.

[54] D. Opaits, G. Neretti, S. Zaidi, M. Shneider, R. Miles, A. Likhanskii, and S. Macheret. DBD Plasma Actuator Driven by a Combination of Low Frequency Bias Voltage and Nanosecond Pulses. In: 46th AIAA Aerospace Sciences Meeting and Exhibit; Reno, USA. AIAA-2008-1372; 2008.

[55] R Dawson, J Little. Characterization of Nanosecond Pulse Driven Dielectric Barrier Discharge Plasma Actuators for Aerodynamic Flow Control. J. Appl. Physics. 2013; 113. DOI: 10.1063/1.4794507.

[56] N. Benard, N. Zouzou, A. Claverie, J. Sotton, and E. Moreau. Optical visualization and electrical characterization of fast-rising pulsed dielectric barrier discharge for airflow control applications. J. Appl. Phys. 2012; 111 (033303). DOI: 10.1063/1.3682568.

[57] R. Sosa, E. Arnaud, E. Memin, G. Artana. Study of the Flow Induced by a Sliding Discharge. IEEE Transactions on Dielectrics and Electrical Insulation. 2009; 16(2). P. 305-311. DOI: 10.1109/TDEI.2009.4815157 .

[58] E. Moreau, C. Louste, and G. Touchard,. Electric wind induced by sliding discharge in air at atmospheric pressure. J. Electrostat. 2008; 66(1-2), 107-114.

[59] A. Santhanakrishnana and J. Jacob,. On Plasma Synthetic Jet Actuators. In: 44th AIAA Aerospace Science Meeting and Exhibit, Reno, NV, 2006. AIAA 2006-317. DOI: 10.2514/6.2006-317, (2006).

[60] N. Bénard, J. Jolibois, E. Moreau, R. Sosa, G. Artana, and G. Touchard. Aerodynamic plasma actuators: A directional micro-jet device. J Thin Solid Films. 2008; 516; 6660-6667. DOI: 10.1016/j.tsf.2007.11.039.

[61] S.H. Kim,W. Hong, and C. Kim. Separation Control Mechanism of Airfoil Using Synthetic Jet. J. Mechanical Science and Technology. 2007; 21, 1367-1375.

[62] N. Bénard, J. Jolibois, G. Touchard, and E. Moreau. A Directional Plasma-Jet Device Generated by Double DBD Actuators - An Active Vortex Generator for Aerodynamic Flow Control. In: 4th Flow Control Conference, Seattle, Washington. 2008. AIAA 2008-3763.

[63] R Joussot, A Leroy, R Weber, H Rabat, S Loyer and D Hong. Plasma morphology and induced airflow characterization of a DBD actuator with serrated electrode. J. Phys. D: Appl. Phys. 2013; 46(125204); 12pp. DOI: 10.1088/0022-3727/46/12/125204.

[64] R J Durscher and S Roy. Three-dimensional flow measurements induced from serpentine plasma actuators in quiescent air. J. Phys. D: Appl. Phys. 2012; 45 (035202);9pp. DOI:10.1088/0022-3727/45/3/035202.

[65] Y.E. Akansu, F. Karakaya, and A. Şanlısoy. Active Control of Flow around NACA 0015 Airfoil by Using DBD Plasma Actuator. EPJ Web of Conferences. 2013; 45(01008): 7pp. DOI: 10.1051/epjconf/20134501008.

[66] C.A. Borghi, A. Cristofolini, G, Neretti, P. Seri, A. Talamelli, and A. Rossetti. Wind Tunnel Experiments on a NACA0015 Airfoil Equipped with Vectorizable

[67] M.L. Post and T.C. Corke. Separation Control on High Angle of Attack Airfoil Using Plasma Actuators. AIAA Journal. 2004; 42: pp. 2177-2184.

[68] C. He, T.C. Corke, and M.P. Patel. Plasma Flaps and Slats: An Application of Weakly Ionized Plasma Actuators. J. Aircraft. 2009; 46; pp. 864-873. DOI: 10.2514/1.38232.

[69] D.P. Rizzetta and M.R. Visbal. Numerical Investigation of Plasma-Based Control for Low-Reynolds-Number Airfoil Flows. AIAA Journal. 2011; 49: pp. 411-425. DOI: 10.2514/1.J050755.

[70] Mark Riherd and Subrata Roy. Serpentine geometry plasma actuators for flow control. J. of Appl. Physics D. 2013; 114(083303). DOI: 10.1063/1.4818622

[71] K. Inaoka, T. Mori, M. Yamaguchi, and M. Senda. Feedback flow control of a low-Re airfoil by flap actuators. Journal of Fluids and Structures. 2015; 58: 319-330. DOI: 10.1016/j.jfluidstructs.2015.08.011.

[72] T. C. Corke, P.O. Bowles, C. He, and E. Matlis. Sensing and control of flow separation using plasma actuators. Phil. Trans. R. Soc. A. 2011; 369: 1459-1475. DOI: 10.1098/rsta.2010.0356

[73] Maden, R. Maduta, S. Jakirlic, S. Grundmann, C. Tropea, and J.K. Eaton. Plasma-Based

[74] S. Grundmann, E. L. Sayles, and J. K. Eaton. Sensitivity of an asymmetric 3D diffuser to plasma-actuator induced inlet condition perturbations. Experiments in Fluids. 2010; 50(1):217–23. DOI: 10.1007/s00348-010-0922-0.

[75] Potočar, E., Širok, B., Hočevar, M., and Eberlinc, M. Control of Separation Flow over a Wind Turbine Blade with Plasma Actuators. Journal of Mechanical Engineering. 2012; 58(1): 37-45. DOI: 10.5545/sv-jme.2011.016.

[76] Yong Li, Xin Zhang, and Xun Huang. The use of plasma actuators for bluff body broadband noise control. Experiments in Fluids. 2010; 49(2); 367-377. DOI: 10.1007/s00348-009-0806-3.

[77] F. O. Thomas, A. Kozlov, and T. C. Corke. Plasma Actuators for Cylinder Flow Control and Noise Reduction. AIAA Journal. 2008; 46(8): pp. 1921-1931.

[78] Xun Huang, and Xin Zhang. Plasma actuators for noise control. International Journal of Aeroacoustics. 2010; 9(4 & 5): pages 679-704.

[79] C. L. Kelley, P. O. Bowles, J. Cooney, C. He, T.C. Corke, B. A. Osborne, J.S. Silkey, and J. Zehnle. Leading-Edge Separation Control Using Alternating-Current and Nanosecond-Pulse Plasma Actuators. AIAA Journal. 2014; 52(9): 1871-1884.doi: 10.2514/1.J052708

[80] Y. Bouremela, J.M. Lia, Z. Zhaob, and M. Debiasia. Effects of AC Dielectric Barrier Discharge Plasma Actuator Location on Flow Separation and Airfoil Performance. Procedia Engineering. 2013; 67: 270-278. DOI: 10.1016/j.proeng.2013.12.026.

[81] Hikaru Aono, Taku Nonomura, Aiko Yakeno, Kozo Fujii and Koichi Okada. Plasma Flow Control Simulation of an Airfoil of Wind Turbine at an Intermediate Reynolds Number. In: 8th Symposium on Flow Manipulation and Active Control: Theory, Experiments and Implementation; July 7-11, 2013; Nevada, USA, . 2013. DOI: 978-0-7918-5555-3.

[82] Miguel Visbal, Datta Gaitonde, and Subrata Roy. Control of Transitional and Turbulent Flows Using Plasma-Based Actuators. In: 36th AIAA Fluid Dynamics Conference and Exhibit, Fluid Dynamics and Co-located Conferences; 5-8 June, 2006; San Francisco, California. 2006. AIAA-2006-3230. DOI: 10.2514/6.2006-3230 .

[83] A. N. M. M. I. Mukut, H. Mizunuma, O. Hiromichi, T. Segawa. Winglet type dielectric barrier discharge plasma actuators: performance characterization and numerical comparison. Procedia Engineering. 2015; 105: 250-257. DOI: http://dx.doi.org/10.1016/j.proeng.2015.05.104.

[84] Xiaohu Zhao, Yinghong LI, Yun Wu, Tao Zhu, Yiwen Li. Numerical Investigation of Flow Separation Control on a Highly Loaded Compressor Cascade by Plasma Aerodynamic Actuation . 2012;25(3):349–360. DOI: 10.1016/S1000-9361(11)60396-8

[85] Philipp C. Dörr, Markus, J. Kloker. Transition Control in a Three-dimensional Boundary-layer Flow Using Plasma Actuators. Procedia IUTAM. 2015;14:469–478. DOI: 10.1016/j.piutam.2015.03.075.

[86] I. Maden, R. Maduta, J. Kriegseis, S. Jakirlič, C. Schwarz, S. Grundmann, C. Tropea. Experimental and computational study of the flow induced by a plasma actuator. International Journal of Heat and Fluid Flow. 2013;41:80–89. DOI: 10.1016/j.ijheatfluidflow.2013.02.013

[87] E. Peersa, Z. Ma, X. Huang . A numerical model of plasma effects in flow control. Physics Letters A. 2010;374(13-14):1501–1504. DOI: 10.1016/j.physleta.2009.08.046

[88] A. Kurz, S. Grundmann, C. Tropea, M. Forte, A. Seraudie, O. Vermeersch, D. Arnal, N. Goldin, R. King. Boundary Layer Transition Control using DBD Plasma Actuators. Journal Aerospace Lab. 2013; 6: AL-06-02. ISSN: 2107-6596.

Chapter 4

Loads Simulator System for Testing and Qualification of Flight Actuators

Nasim Ullah

Additional information is available at the end of the chapter

http://dx.doi.org/10.5772/62710

Abstract

The flight actuation system plays important role in the accurate guidance of the flight vehicles. The actuators driving the control surfaces are aerodynamically loaded during flight. The design, testing and selection process of the flight actuators play important role to ensure the stable and safe flight. Since a reliable flight actuation system can ensure appropriate guidance, the importance of qualification process cannot be neglected. Qualification of the actuators through field trials is a very costly and time-consuming process. The testing process using real flights takes more time and is costly. For ground testing, aerodynamic loading systems are used. The aerodynamic loading system is ground-based hardware in the loop (HWIL) simulator that can be used for exerting aerodynamic loads on actuation system of flight vehicles in real-time experiment. The actuation system under test is directly connected to the loading motor through a stiff shaft and the aerodynamics loading is applied in real time according to the flight trajectory generated by a flight computer.

This chapter is preliminarily focused on the basic working principle, mathematical modelling and torque/force control system design of the load simulator system. As a case study, the response and dynamics of the electrical aerodynamic loading system are analysed using mathematical modelling concept. The dynamic model is discussed and adaptive fuzzy sliding mode control techniques are introduced to ensure the high-performance torque loop of the aerodynamic loading systems.

Keywords: Ground-based testing, aerodynamic load simulators, flight actuators, mathematical modelling, modern control

1. Introduction

In a flight control system, aerodynamic loads are introduced during flight at the control surfaces or fins of flight vehicle as a function of air density, fin angle, etc. The fin control system must respond to these loads to maintain the accuracy of flight control system [1–3]. Since load simulators are used for the qualification of flight actuation system, the control performance of load simulators is vital [4, 5]. Before the invention of the motorized load simulators, the mechanical springs or torsion bars were utilized for the testing process. Conventional testing has several disadvantages, for example, the loading devices are not adaptable to varying loads. In modern ages, the testing and qualification of flight actuators are done through a hardware in the loop simulator system. In order to reduce cost and time, hardware in the loop simulators are utilized in which a loading motor is used as the aerodynamic loading device [6–9]. There are three types of load simulators depending on the type of applications and loading requirements [2–6]. Electro-hydraulic load simulators (EHLS) are widely used as hardware-in-the-loop-simulation (HILS) system in flight control applications that can simulate the air load executed on flight actuation systems. The EHLS can simulate rapid and large torque/force loads with big volume. Some advantages and disadvantages are listed. The hydraulic load simulators move fast and have rapid steering. The natural frequency of the electro-hydraulic servo valve is generally above 100 HZ, so the hydraulic load simulator's frequency response is very fast and smooth. The low-speed performance of electro-hydraulic actuator is good. The load simulators can be operated in wide speed range. On other hand, due to big size and volume, maintenance cost is high. Temperature effects can cause degradation of performance. At low temperature the friction loss is high, and at high temperatures the leakage phenomenon is significant. Leakage of hydraulic fluid or oil is a major drawback. It causes the environmental pollution. This can easily cause a fire. The oil contamination causes system blockage and system failures. Another type is the pneumatic load simulator that is a typical hardware in loop (HIL) system used to apply medium-range torque/force loads on the actuator under test. The pneumatic load simulator can track the fast deployed load with compressibility. Pneumatic systems are often used instead because pneumatic pressure is usually cheaper to obtain, especially since most industrial facilities already have compressed air available. The pneumatic load simulators use air or any other compressed gas to transmit power. It is hard to obtain smooth performance due to large compressibility factor, especially at low speeds. Similarly, temperature changes can degrade the performance of pneumatic load simulators. The pneumatic load simulators are smaller in size and volume as compared to hydraulic load simulators, so the maintenance cost is lower. The electrical load simulator is an important ground-based hardware tester simulator used for qualification of flight actuation system, such as the aircraft control surfaces, ship steering system, robotics arm, undercarriage of the plane and high-speed elevator system. The electrical load simulators are used to apply medium and small loads on actuators under test. Some advantages of electrical load simulator are its small size and volume, low installation and maintenance cost, less sensitive to temperature changes as compared to hydraulic and pneumatic load simulators and high performance. There are no environmental hazards associated with electrical load simulators.

http://dx.doi.org/10.5772/62710

2. Working of load simulator system

Laboratory testing and qualification of flight actuators play vital role for ensuring the design of reliable flight control with cost-effective solutions. The hardware diagram given in **Figure 1** is the minimal setup that is used in the laboratory testing of the flight actuators. The flight actuator is directly connected to the loading motor through a stiff shaft. The direct mechanical connection allows the application of the loading torque generated at the output shaft of loading motor in a close loop system. The Autopilot controller is used to set the reference position trajectory of the flight actuator. An important block of the system is the aerodynamic load calculator system, and the inputs to this block are the reference position command of the actuator system, air speed, Mach number and angle of attack. The output of the load calculator block serves as reference command to the loading motor. Thus, the torque load loop is closed using the feedback signal from torque sensor. The advantage of using the proposed setup provides the designer a freedom for applying real-time loads simulated in the laboratory with minimum cost.

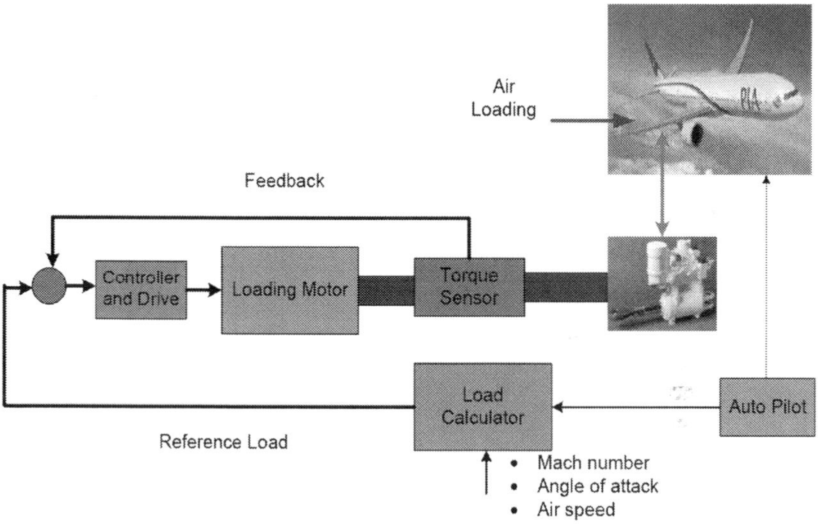

Figure 1. Block diagram of electrical load simulator.

3. Actuator testing: a case study

The flight actuator under test is a brushless DC motor. A permanent magnet synchronous motor (PMSM) is used as the loading motor. For simplicity, the load calculator is used in proportional mode. In proportional mode, the output of the load calculator is directly proportional to the reference command of the flight actuator. Let θ_a represents the reference command of flight actuator, then the output of the load calculator T_r is $C\theta_a$ where "C" represents the proportionality constant. The parameter T_r is set as reference command of the torque loading motor. From **Figure 2**, the reference command signal of the flight actuator $\theta_a$is constructed from a real test scenario of a flight vehicle.

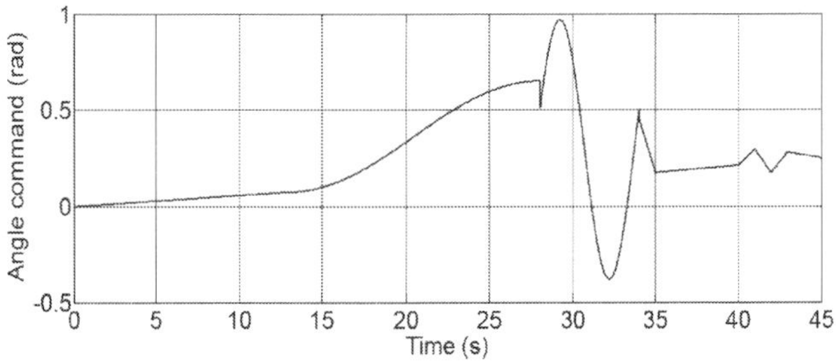

Figure 2. Reference signal command of flight actuator.

Before discussing the control problem, basic understandings of system dynamics is a pre requisite. A PMSM is used as a loading motor. The voltage and torque balance equations of PMSM loading motor are written as

$$
\begin{aligned}
u_d &= i_d R_s + L_{sd}\frac{di_d}{dt} - PL_s i_q w_m \\
u_q &= i_q R_s + L_{sq}\frac{di_q}{dt} + PL_{sq} i_d w_m + P\psi_m w_m \\
T_e &= \frac{3P}{2}[\psi_m i_q + \left(L_{sd} - L_{sq}\right) i_d i_q = J\frac{dw_m}{dt} + Bw_m + T_f + T_L
\end{aligned}
\tag{1}
$$

In Eq. (1) $[i_d\, i_q]$ is the d-axis and q-axis current vector, $[u_d\, u_q]$ represents d-axis and q-axis voltage vector, w_m is the angular velocity of torque motor, $[L_{sq}\ L_{sd}\ R_s]$ represents inductances and resistances of ELS motor,$[P\, ,\psi_m]$ represents the number of pole pairs and magnetic flux of rotor, $[J, B]$ represents inertia and damping coefficient, $k_t = \frac{3P}{2}\psi_m$is torque constant, $k_b = P\psi_m$ is back emf constant, and $[T_e, T_f, T_L]$ is the electromagnetic torque, friction torque and output load torque. Assuming that inertia and damping coefficient of torque sensor is very small, thus the dynamics can be written as

$$T_L = K_s\left(\theta_m - \theta_a\right) \tag{2}$$

Here, $[\theta_m\, \theta_a]$ represents the angular position of loading motor and actuator, K_s is the total stiffness of torque sensor and connecting shaft. To achieve largest torque operation and to eliminate coupling effect between speed and currents, we set d-axis reference current i_d' equal to zero. Considering the effect of PWM driver and the current feedback, Eq. (2) can be written as given in [4]. For simplicity, we set $L_{sq} = L$ and $R_s = R$

$$\frac{di_q}{dt} = -\frac{R}{L}i_q - \frac{K_b}{L}w_m + \frac{K_v K_i}{L} \tag{3}$$

Here, k_i is the current controller gain. Current feedback is assumed to be unity. Assuming that load and friction torque are zero and taking Laplace transform of Eq. (3) and eliminating $I_{(s)}$, we get the transfer function from input voltage to output position as

$$\frac{\theta_m}{u_q} = \frac{\frac{K_v K_i}{K_b}}{S\left(t_m t_e S^2 + t_m S + 1\right)} \tag{4}$$

In Eq. (4), $t_m = \frac{RJ}{k_t k_b}$ is the electromechanical time constant, $t_e = \frac{L}{R}$ is the electromagnetic time constant. Replace Eq. (2) into Eq. (2), the simplified relation can be written as

$$\begin{aligned} T_L &= \frac{k_v k_i k_t u - s[(Ls+R)(Js+B)+k_b k_t]K_s \theta_a}{D(s)} \\ D(s) &= s\left[(Ls+R)(Js+B)+K_b K_t\right] + T(Ls+R) \end{aligned} \tag{5}$$

From the numerator of Eq. (5), it is concluded that extra torque is caused by the effect ofθ_a, θ_a and $\ddot{\theta}_a$. If the reference input command of loading torque motor is zero, i.e.$u = 0$, then Eq. (5) is reduced to the following simplified relation

$$T_L = \frac{-s[(Ls+R)(Js+B)+k_b k_t]K_s \theta_a}{D(s)} \tag{6}$$

From Eq. (6), it is concluded that extra torque is acting on the loading torque motor even if the reference input command u is zero. Extra torque is a function of the acceleration and velocity components of the actuators movement. After some simplifications, the state equation representation of electrical load simulator is written as

$$\ddot{T}_L = -a\dot{T}_L + bu - cf\left(T_{extra}, T_f, T_L\right) \tag{7}$$

In Eq. (7) the parameters are defined as

$$a = \frac{k_t k_b}{JR} + \frac{B}{J}$$
$$b = \frac{K_s k_t}{JR}$$
$$c = \frac{K_s}{J}$$
$$f\left(T_{extra}, T_f\right) = \frac{K_s}{J}\left(T_{sft} + T_f\right)$$

In Eq. (7) the nonlinear friction is represented using LuGre model, which is written as

$$\begin{aligned} T_f &= a_0 Z + a_1 \dot{Z} + a_2 \\ \dot{Z} &= v - \frac{a_0 \mid v \mid}{g(v)} \\ g(v) &= f_c + (f_c - f_s)e^{[\frac{v}{v_s}]^2} v \end{aligned} \tag{8}$$

In Eq. (8) the parameter $g(v)$ is the Stribeck effect, v_s is the Stribeck velocity, f_c is coulomb friction, f_s is static friction, z is the average bristle defection, a_0 is the stiffness of the bristles, a_1 is the damping term and a_2 is the viscous friction coefficient. Now to realistically apply the loading torque on flight actuator, a feedback torque control system plays vital role. In this study, adaptive fuzzy sliding mode control system is used to formulate the torque control system.

3.1. Adaptive fuzzy sliding mode control for electrical load simulator system

Sliding mode is a robust control method which has been widely applied to many nonlinear systems [10–15]. This section provides an overview of derivations of torque control system for electrical load simulator's system. Let T_L be the output load torque and T_r be the desired torque signal, we define tracking error vector as

$$\begin{bmatrix} e = T_L - T_r \\ \dot{e} = \dot{T}_L - \dot{T}_r \\ \ddot{e} = \ddot{T}_L - \ddot{T}_r \end{bmatrix} \tag{9}$$

Error surface vector is defined as

$$\begin{bmatrix} s = \dot{e} + \lambda e \\ \dot{s} = \ddot{e} + \lambda \dot{e} \end{bmatrix} \tag{10}$$

Assuming that the nominal parameters of the system are known, then by combining Eqs. (7), (9) and (10) yields

$$\dot{s} = \left(-a\dot{T}_L + bu - f\left(T_{extra}, T_f\right)\right) - \ddot{T}_r + \lambda \dot{e} \tag{11}$$

The control law is given by

$$u = \frac{1}{b}\left(a\dot{T}_L + \hat{f}\left(T_{extra}, T_f | \theta\right) + \ddot{T}_r - \lambda \dot{e}\right) - \frac{1}{b}\left(K_d.s - w.sgn(s)\right) \tag{12}$$

From Eq. (12), it can be analyzed that the total control effort u is the sum of three terms

$$u = u_T + u_f + u_{extra} \tag{13}$$

Here, u_T is the control effort for torque tracking, u_f is the friction compensation control and u_{extra} is the extra torque compensation control. The unknown function $\tilde{f}(T_{extra,} T_f \mid \theta)$is the estimated output of fuzzy logic for friction and extra torque.

3.1.1. *Stability analysis*

To prove stability of the closed loop, the Lyapunov function is chosen as

$$\begin{gathered} V = \frac{1}{2}\left(s^2 + \sum_{i=1}^{n} \eta_i \tilde{\theta}^2{}_i\right) \\ \dot{V} = s\dot{s} + \sum_{i=1}^{n} \eta_i \tilde{\theta}_i \dot{\tilde{\theta}}_i \end{gathered} \tag{14}$$

Here$\tilde{\theta}_i = \hat{\theta}_i - \theta_i$. Combine Eq. (11) and Eq. (14)

$$\dot{V} = s\left(\left(-a\dot{T}_L + bu - f(T_{extra}, T_f)\right) - \ddot{T}_n\right) + \sum_{i=1}^{n} \eta_i \tilde{\theta}_i \dot{\tilde{\theta}}_i \tag{15}$$

Define $\dot{T}_n = -\ddot{T}_r + \lambda\dot{e}$ and combine Eq. (12) into Eq. (15)

$$\begin{aligned} \dot{V} = s\Big(& -a\dot{T}_L + a\dot{T}_L + \hat{f}(T_{extra}, T_f)\Big|\theta + \ddot{T}_n - K_d s - w.sgn(s) \\ & - f(T_{extra}, T_f) - \ddot{T}_n\Big) + \sum_{i=1}^{n} \eta_i \theta_i \dot{\theta}_i \end{aligned} \tag{16}$$

The fuzzy approximation error is defined as [3]

$$\begin{aligned} e_f &= f\left(T_{extra}, T_f\right) - \tilde{f}\left(T_{extra}, T_f\right) | \theta^* \\ \theta_i \xi_i\left(\dot{\theta}\right) &= \hat{f}\left(T_{extra}, T_f\right) / \theta - \tilde{f}\left(T_{extra}, T_f\right) | \theta^* \end{aligned} \tag{17}$$

Combining Eq. (16) and Eq. (17) yields [16]

$$\dot{V} = s(\hat{f}(T_{extra}, T_f) / \theta - f(T_{extra}, T_f) - K_d s - w.sgn(s)) + \sum_{i=1}^{n} \eta_i \theta_i \dot{\theta}_i \tag{18}$$

$$\begin{aligned} \dot{V} = s\Big(& \hat{f}(T_{extra}, T_f)\Big/\theta - \tilde{f}(T_{extra}, T_f)\Big|\theta^* - ((f(T_{extra}, T_f) - \tilde{f}(T_{extra}, T_f) \,|\, \theta^*)) \\ & - K_d s - w.\operatorname{sgn}(s)\Big) + \sum_{i=1}^{n} \eta_i \theta_i \dot{\theta}_i \end{aligned} \tag{19}$$

Using Eq. (19) the following adaptive law is derived

$$\dot{\theta}_i = -\eta_i^{-1} s_i \xi_i\left(\theta, \dot{\theta}\right) \tag{20}$$

By replacing Eq. (20) in Eq. (19) and simplifying

$$\dot{V} = s\left(-e_f - K_d s - w.\operatorname{sgn}(s)\right) \tag{21}$$

It is assumed that ideally fuzzy compensating error e_f is approaching zero, and by choosing $K_d > 0$ it can be shown that

$$\dot{V} = -sK_d s \leq 0 \tag{22}$$

3.1.2. Results and discussion

For simulations and validity of the proposed control scheme, the following parameters are used. Total inertia of the system is given as $J = 0.04Kg/m^2$, resistance $R = 7.5\Omega$, inductance $L = 1mH$, motor torque constant $k_t = 5.7325Nm/A$, back emf constant $k_b = 5.7325Nm/V$, viscous coefficient $B = 0.244Nm/rad/s$, torque sensor stiffness $K_s = 950Nm/rad$, static friction $f_s = 3Nm$, coulomb friction $f_c = 2.7Nm$, $\sigma_0 = 260Nm/rad$, $\sigma_1 = 2.5Nm - s/rad$, $\sigma_0 = 0.022Nm - s/rad$ and Stribeck velocity $\alpha = 0.001rad/s$. The parameters of the controller are given as fuzzy learning rate $\eta_i = 0.0001$, amplifier gain $k_u = 10$, $k_d = 10$, $w = 1.5$, $\lambda = 15$.

3.1.3. Loading motor torque tracking performance

The testing of actuator is performed under the loading torque $T_r = 16\theta_a$ where "*C = 10*". From **Figure 3**, it is concluded that the output torque applied by the loading motor is exactly the same as the reference loading torque.

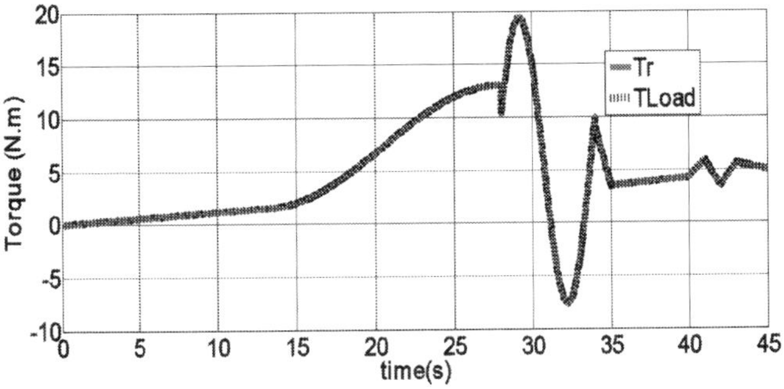

Figure 3. Loading motor closed loop performance.

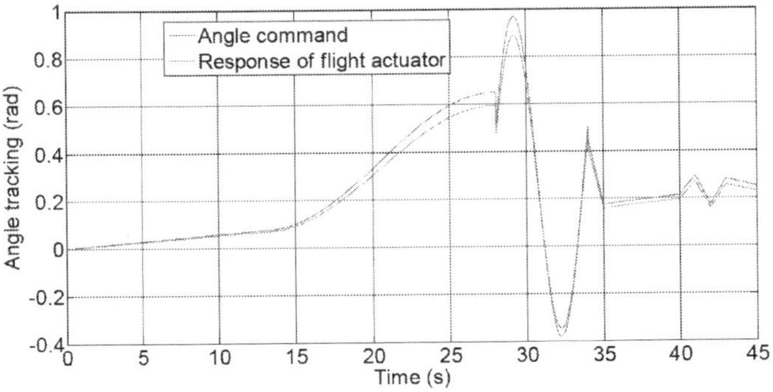

Figure 4. Flight actuator closed loop performance with aerodynamics load.

3.1.4. Flight actuator angle tracking performance under load

Figure 4 presents the testing results and the qualification of the autopilots of the flight actuators under the aerodynamic loading shown in **Figure 3**, which is mechanically transmitted from loading motor. From the results provided, it is concluded that the flight actuator under can withstand the non-linear profile of the aerodynamic load supplied. Moreover, the autopilot position controller is also robust and the position tracking errors are small enough.

4. Conclusion

This chapter covers the basic working principle of the load simulator system for testing of the flight actuators. As a case study, a real test data was used as reference command to the flight actuators and the load calculator unit. The output of the load calculator system provides the reference loading torque command for the loading motor, which is working in closed loop. From the results presented in this chapter, it is concluded that the proposed hardware setup is feasible to be utilized for cost-effective testing and qualification of the flight actuation system.

Author details

Nasim Ullah

Address all correspondence to: nasimullah@cusit.edu.pk

City University of Science and IT, Peshawar, Pakistan

References

[1] Zongxia J., Chenggong L., and Zhiting R.. The extraneous torque and compensation control on the electric load simulator. In: International Symposium on Instrumentation and Control Technology; 2003. pp. 723–727.

[2] Wang X., Wang S., and Yao B.. Adaptive robust control of linear electrical loading system with dynamic friction compensation. IEEE/ASME International Conference on Adavanced Intelligent Mechatronics, 2010: pp. 908–913.

[3] Nasimullah S. Wang, and Aslam J.. Adative robust control of electrical load simulator based on fuzzy logic compensation. International Conference on Fluid Power and Mechatronics, 2011: pp. 861–867.

[4] Wang X. and Wang S.. Electrical load simulator based on velocity-loop compensation and improved fuzzy-PID. IEEE International Symposium on Industrial, 2009: pp. 238–243.

[5] Chen K., Wang J., and Yan J.. Experiment and study of electric loading simulator for linear rudder. Electronic Engineering, 2008, 1: 4798–4802.

[6] Changhua L.. Design of electric loading system in flight simulator based on PIDNN [A]. International Conference on Mechatronic Science, Electric Engineering and Computer, 2011: pp. 2623–2626.

[7] Liu S., Wang M., Tian K., and Wang Y.. Research on loading simulation of DC torque motor for electrical load simulator. IEEE Conference on Industrial Electronics and Applications [C], 2008: pp. 1146–1150.

[8] Huang Y., Chen K., and .Wei J. Robust controller design and experiment for electric load simulator. International Advanced Computer Theory and Engineering, 2010: pp. 236–240.

[9] Truong D. Q., Kwan A. K., and Il Yoon J.. A study on force control of electric-hydraulic load simulator using an online tuning quantitative feedback theory [A]. International Conference on Control, Automation and Systems [C], 2008, pp. 2622–2662.

[10] Me H., Hong S., and Ik S.. Mechatronics Precise friction control for the nonlinear friction system using the friction state observer and sliding mode control with recurrent fuzzy neural networks [J]. Mechatronics, 2009, 19(6): 805–815.

[11] Erfanian V. and Kabganian M.. Adaptive trajectory control and friction compensation of a flexible-link robot [J]. Scientific Research and Essay, 2009, 4(4): 239–248.

[12] Angue-mintsa H. and Belleau C.. Adaptive Position Control of an Electrohydraulic Servo System With Load Disturbance Rejection and Friction Compensation [J]. Journal of Dyanmic systems, Measurment and Control, 2011, 133(6): 1–8.

[13] Lu L., Yao B., Wang Q., and Chen Z.. Adaptive robust control of linear motor systems with dynamic friction compensation using modified LuGre model [A]. International Conference on Advanced Intelligent Mechatronics [C], 2008: pp. 961–966.

[14] Messaoudi M., Sbita L., and Abdelkrim M. N.. A robust nonlinear observer for states and parameters estimation and on-line adaptation of rotor time constant in sensorless induction motor drives. International Journal of physical sciences, 2007, 2(8): 217–225.

[15] Ullah N. , Shaoping W., Khattak MI. and Shafi M.. Fractional order adaptive fuzzy sliding mode controller for a position servo system subjected to aerodynamic loading and nonlinearities [J]. Aerospace Science and Tehcnology, 2015, 43: 381–387.

[16] Ullah N., Han S., and Khattak MI.. Adaptive fuzzy fractional-order sliding mode controller for a class of dynamical systems with uncertainty. Transactions of the Institute of Measurement and Control, 2015, June 3, 20150142331215587042

Chapter 5

Morphing Technologies: Adaptive Ailerons

Ignazio Dimino, Gianluca Amendola,
Francesco Amoroso, Rosario Pecora and
Antonio Concilio

Additional information is available at the end of the chapter

http://dx.doi.org/10.5772/63645

Abstract

European Union is involving increasing amount of resources on research projects that will dramatically change the costs of building and operating aircraft in the near future. Morphing structures are a key to turn current airplanes to more efficient and versatile means of transport, operating into a wider range of flight conditions.

The concept of morphing may aim at a large number of targets, and its assessment strongly depends on the final objectives and the components where it has to be deployed. Maneuver, takeoff, landing, cruise conditions, just to cite few and very general examples, have all their own peculiarities that drive the specifications the wing shape change has to suit on.

In general, an adaptive structure ensures a controlled and fully reversible transition from a baseline shape to a set of different configurations, each capable of withstanding the relative external loads. The level of complexity of morphing structures naturally increases as a consequence of the augmented functionality of the reference system. Actuation mechanisms constitute a very crucial aspect for adaptive structures design because has to comply variable wing shapes with associated loads and ensure the prescribed geometrical envelope.

This chapter provides a presentation of the state of the art, technical requirements, and future perspectives of morphing ailerons. It addresses morphing aircraft component architecture and design with a specific focus on the structural actuator system integration. The approach, including underlying concepts and analytical formulations, combines methodologies and tools required to develop innovative air vehicles. Aileron is a very delicate region, where aeroelastic phenomena may be very important because of the very reduced local stiffness and the complex aerodynamics, typical of the wingtip zone. On the other side, this wing segment showed to be the one where higher cruise benefits could be achieved by local camber variations. This target was achieved while keeping the typical maneuver functions.

Keywords: morphing, actuation systems, distributed actuation, wind tunnel tests, aileron, lift control

1. Introduction

Men desired to flight since very ancient times being inspired by bird's capability to dominate sky. Nature offers a rich seam of inspiration for a new generation of morphing wing design across a wide range of scales of interest to engineers going from the biggest birds to the smallest insect. For example, birds achieve their wing morphing capability using flexible lifting surfaces, stiffened by hollow bones attached to strong muscle. All the flying creatures of the world show an inherent capacity to adapt, in a fraction of a second, their wing shape as the flight condition changes. A very interesting example may be represented by the perching sequence of an eagle. As reported in [1], birds accomplish changes in wingspan and area by firstly flexing their wings, and then adopting a characteristic M-shape planform with the inner wing section sweeps forward, and the outer section sweeps backwards.

It is noteworthy that "inspiration from nature" is the keywords that lie behind any morphing idea. Many researchers and engineers around the world have been inspired by the multitasking flight capabilities of birds, which tend to cover a broad range of mission phases ranging from slow, near-hover flight to aggressive dives, in order to develop innovative methodologies involved to resolve many technological problems. Just only observing birds and other flying creature wings, it is possible to appreciate the complexity of such systems showing intrinsic capacities to adapt instinctively and immediately to the environment. In particular, birds are able to articulate their wings in a craning motion to vary the dihedral or sweep angles [1], wing area, wing planform, wingspan, and other parameters. These changes allow the bird to quickly adapt between soaring, cruising, and descending flight [1].

The morphing idea was well known by the engineering since the beginning of aviation such as the Wright brothers who built the "first heavier than air aircraft with engine" with twisted wing for roll control. Despite the past century of innovation in aircraft technology, the versatility of modern aircraft remains far worse than airborne biological counterparts. The shape modification accomplished by birds stands as one of the few examples of true morphing. As such, the aircraft engineers worldwide are devoting extensive effort to integrate these concepts in advanced mechanical systems in order to bring morphing technology to the readiness level of a flight vehicle. The key purpose is to realize an innovative device capable to adapt itself to the external environment conditions, by exhibiting an intrinsic multidisciplinary attitude involving structures, actuation, sensing, and control. In recent years, European community funded many research program involved to improve the morphing structures technology readiness level. SARISTU [2] (acronym of Smart Intelligent Aircraft Structures) was probably the most advanced large-scale integrating project on morphing structures, coordinated by Airbus, aiming at achieving reductions in aircraft weight and operational costs, as well as an improvement in the flight profile specifically related to aerodynamic performance. Ended in 2015, the project consisted of a joint integration of different conformal morphing concepts in a laminar wing with the aim to improve aircraft performance through a 6% drag reduction inside the lift coefficient range usually devoted to cruise, with a positive effect on fuel consumption. The final product of the project was the first full-scale completely morphing wing tip prototype, ever assembled in Europe, at Finmeccanica Headquarters (Pomigliano,

http://dx.doi.org/10.5772/63645

Italy), **Figure 1**. The innovative seamless morphing wing incorporates a gapless morphing leading edge, a morphing trailing edge, and an adaptive winglet.

Figure 1. Assembly of the SARISTU morphing wing consisting of different morphing devices [2].

Morphing technology is now approaching the high maturity practices for the integration on real aircraft. *How to adapt* is a problem regarding sensing, actuation, and control laws, which are very critical. Hence, although an animal's wings may be able to change shape in a complex manner, the total number of independently controlled degrees of freedom may not be high. This indicates that a smart structure is built upon relatively simple principles. It will be actuated in one point and, by means of movable structural elements with limited DOF; the movement is transmitted to the whole structure so that the wing will be built to adapt at loading rather than to resist it.

1.1. Actuation systems for morphing applications

The state of the art of high-lift actuation systems of aircraft control surfaces predominantly consists of mechanical transmission shafts moved by rotary or linear hydraulic actuators with common control valves. These architectures assure a synchronous, safe, and reliable deployment of all HLD (High Lift Device) but with limited flexibility [3]. The main functionality of the high-lift devices is to provide lift increment at low-speed condition (take/off and landing) so that the clean wing is optimized for the cruise speed regime. There are a lot of HLD on wing aircraft such as plain flaps to fowler flaps with single, double, and even the most complex triple slots (Boeing 747). The design and optimization of high-lift systems are one of the most complex tasks in aircraft design. It involves a close coupling of aerodynamics, structures, and kinematics. The evolutionary trend of the HLD has been strongly driven by the dramatic improvement in aerodynamic tools optimization and in computational systems for complex structure simulations (multi-body kinematics). At the early stage, the research of aerodynamics high-lift performance (C_{Lmax}) was achieved by means of multi-slotted experimentally validated two-dimensional flap design. These systems allowed to achieve satisfactory performance with penalties in structural complexity and weight and, therefore, in costs that were not sustainable in the current applications. Later on, the improvement in computation fluid dynamics has permitted to carefully optimize flap systems in two-dimensional flow with a clear advantage for fowler mechanism that allowed to reach higher values of maximum lift due to the effect of an increased lifting surface. Such fowler mechanism, on the other side, required even more complex kinematic actuation system due to a combination of two movements: one translation and a rotation. The fowler flap deployment mechanisms were designed using linear or curved tracks in conjunction with revolute joint for the rotation, but unfortunately, the high-lift values achieved were compensated by the relatively high weight penalties introduced by such systems. The reason for such high weight drawbacks was due to very intensive loads to be withstood by track bearings with also subsequent high maintenance costs. More recently, the research for aerodynamic efficiency and reduced weight penalties and complexity has been fostered by large utilization of multi-body system optimization that permitted the development of lighter and more efficient kinematic mechanism such as multi-link system. Such devices permit to match even very complex aerodynamic requirements with relatively structurally efficient system. As a matter of fact, today it seems very difficult to further improve in terms of an optimum balance among aerodynamic, structural weight, and complexity in the

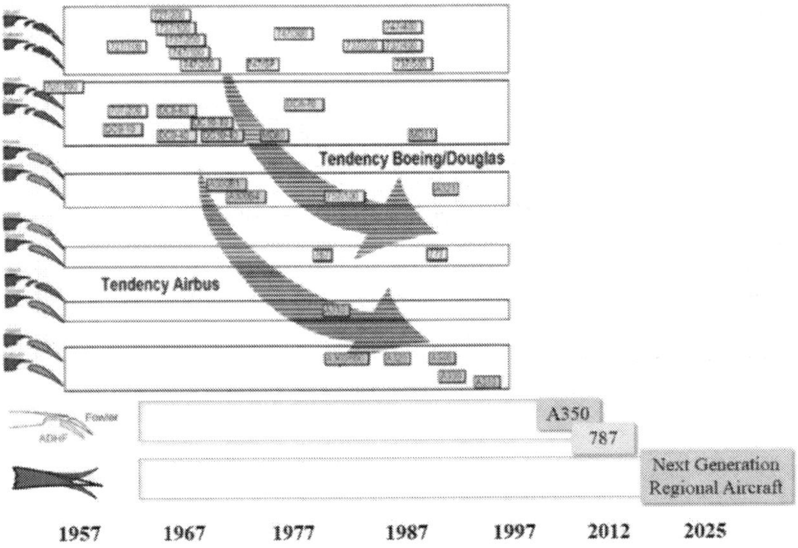

Figure 2. Evolutionary trend in high-lift systems [4].

current system, namely A350 or Boeing 767, this appears evident by the flattening of the curve in **Figure 2**.

From the previous graph, it is evident that today's high-lift system are moving toward the development of innovative mechanisms with continuous curvatures, leading to the removal of gaps in order to obtain the same performance with the less deflections. In other words, this means implementing morphing concepts, as highlighted in the graph reported in **Figure 3**.

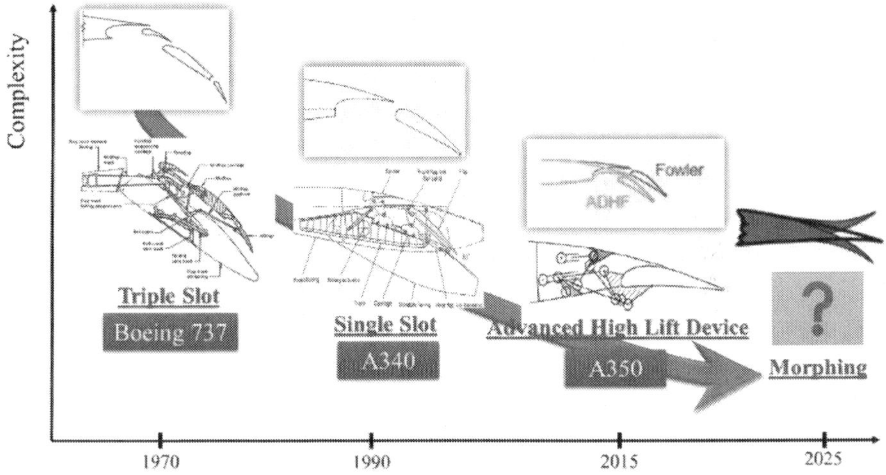

Figure 3. Simplification of the high-lift actuation systems over the last few decades.

Additionally, flap mechanisms must be reliable and fail-safe. In order to not violate safety needs, the driving idea is to elude a multitude of links and joints in series, where high load concentrations are located; because the failure of any one of which could either locks up the flap, make it collapse. There are many type of flap mechanism that are largely investigated in [4, 5]. The actuation scheme of the Airbus A340 and its extraction device are depicted in **Figures 4** and **5**. The central hydraulic power control unit (PCU) supplies the power necessary to deflect the flap panels on each wing. A mechanical transmission shaft transmits the mechanical power to the rotary actuators, which move the flaps on the tracks. This shaft system consists of gearboxes necessary for larger direction changes as well as system torque limiters, wing tip brakes, universal joints, plunging joints, and spline joints to accommodate wing bending and temperature effects. The high-lift system is controlled and monitored by two slat-flap control computers (SFCC) using sensor information from several analogue and discrete sensors. This type of mechanical transmission shaft system consists of a high number of components with different part numbers and requires high design-engineering and installation effort.

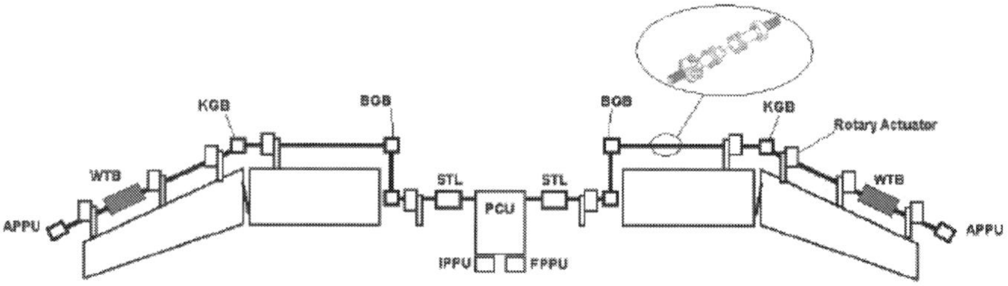

Figure 4. Global scheme of the inboard and outboard A340 flap actuation system [3].

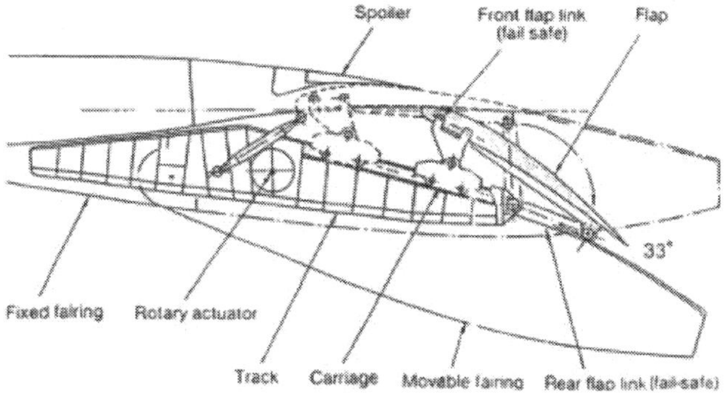

Figure 5. A340 flap mechanism based on the link/track architecture [5].

In contrast to the previous mechanism, the flap deployment system of the Boeing 767 (**Figure 6**) is based on a limited number of links in order to create an articulated quadrilateral or more complex hexagonal chain.

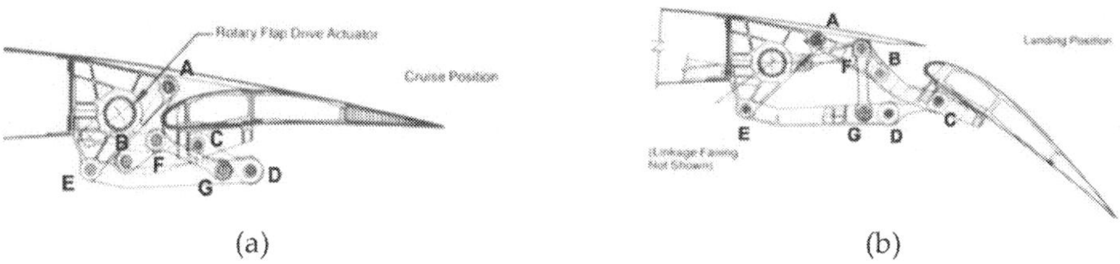

Figure 6. Boeing 767 flap system: cruise position (a) and landing configuration (b) [5].

Recent development programs at Airbus and Boeing extend the functional capabilities of the flap systems. The A350 XWB as well as the B787 high-lift systems design will incorporate additional functionalities that provide aircraft performance optimization. Additional functionality is achieved with an evolution of the traditional mechanical transmission shaft system and additional active components [6]. The A350's flaps are a very simple "drop-hinge" design with a single slot between the trailing edge of the spoiler and the leading edge of the flap. As the flap extends, the spoilers deflect downwards to control the gap and optimize the high-lift performance of flap. It constitutes a multipurposes high-lift system with augmented functionalities, and furthermore, it is a lightweight structures thanks to its low complexity link-based kinematic. This can be summarized in the next **Figures 7** and **8**.

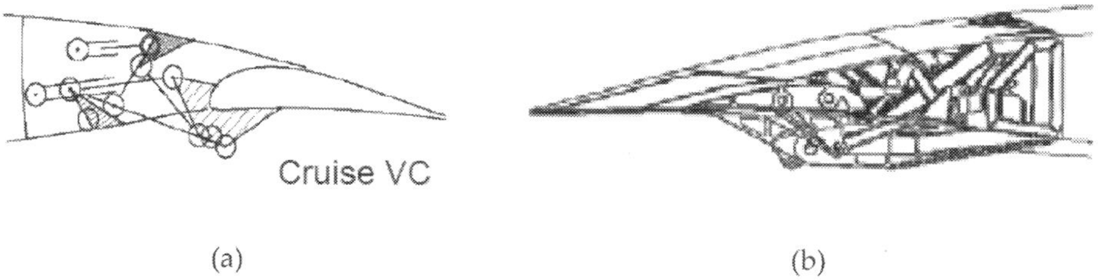

Figure 7. A350 XWB (Extra-Wing Body) flap in cruise condition [6].

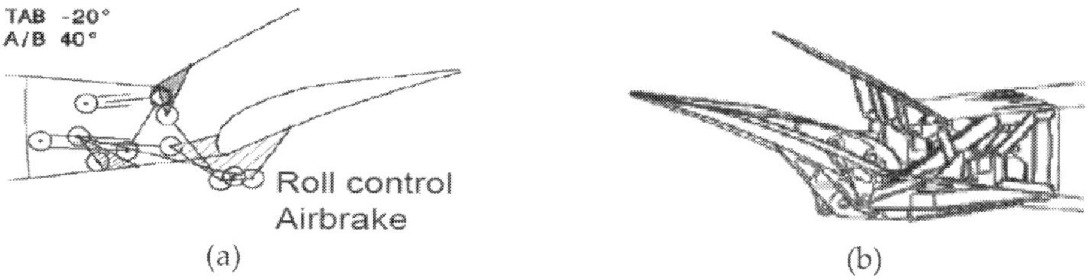

Figure 8. A350 XWB (Extra-Wing Body) with A/B and tab deflection for roll control maneuver [6].

Moreover, for the first time, the flap system will have the both the capability for differential inner and outer settings as well as a variable camber function. The design is composed of a gearbox with a motor installed between the outer and inner flap that enables a differential control of the relative angle in order to shift inboard the resultant lift for a less bending moment. Furthermore, both inner and outer flaps can be moved together during the cruise to optimize the wing's camber for each phase of the flight and use the polar of drag to its most efficient configuration [6].

It remains to discuss if, as the complexity level of the actuation mechanism seems to reduce, the promise of morphing aircraft will become feasible within the next few years. If so, how morphing devices will be actuated?

The next technological challenge, envisaged in the context of more or all-electric aircraft, will be to replace the heavy conventional hydraulic actuators with a distributed spanwise arrangement of smaller electromechanical actuators (EMAs). This will bring several benefits at the aircraft level: firstly, fuel savings. Additionally, a full electrical system reduces classical drawbacks of hydraulic systems and overall complexity, yielding also weight (-15%) and maintenance benefits. Lack of supply buses, improved torque control, enhanced efficiency, removal of fluid losses and flammable fluids are only some of the benefits that can be achieved. On the other hand, a general limit of electro-mechanic actuators is the possibility of jamming failures that can lead to critical aircraft failure conditions. **Figure 9** shows a practical comparison between the aircraft torque shaft configuration and a distributed actuation arrangement suitable for a morphing trailing edge device.

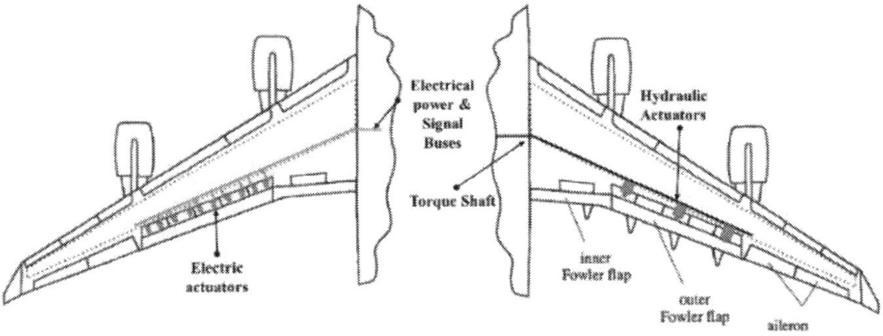

Figure 9. Distributed concept versus concentrated actuation concept.

The simultaneous need for monitoring target morphed shapes, actuation forces, and flight controls along with the counter-effects of aerodynamic loads under aircraft operating conditions suggests the use of a ground-based engineering tool for the physical integration of systems. The most suitable to optimize and validate such systems including electromechanical component such as actuators and flight controls is the "Iron Bird." The basic scheme of an Iron Bird suitable for the integration of different morphing systems is depicted in **Figure 10**. It includes different morphing devices installed on an aeroelastically reasonable aircraft wing box as well as the basic equipment needed to carry out "hardware in the loop simulations." Such a concept may be used to demonstrate advanced control technologies in a modular multi-level design that provides the robustness and the flexibility of a real aircraft integration. Manufacturing, assembly, and integration issues including electrical and flight control may be extensively addressed in relation to the actual configuration of the aircraft. It is the perfect tool to confirm the characteristics of all system components or to discover an incompatibility that may require modifications during early development stages, and thereby, it accelerates the transition to test in a relevant environment. Additionally, failures and mitigation actions introduced in the systems can be studied in full detail and recorded for analysis using such a dedicated testbed.

http://dx.doi.org/10.5772/63645

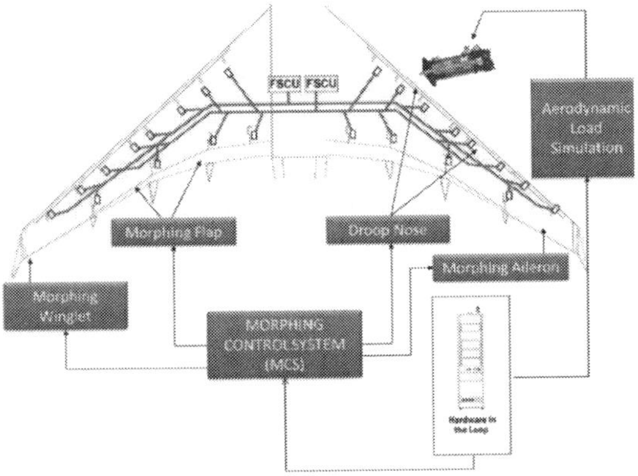

Figure 10. Representative scheme of an Iron Bird tool suitable for testing morphing devices.

The "Iron Bird" for testing morphing wing architectures enables test engineers to evaluate the real-time capabilities of morphing devices with the purpose of:

- demonstrating maturity, reliability, and integrated performance of morphing devices that otherwise could only be achieved with more expensive costly and less safe methods such as wind tunnel tests or flight tests;
- optimizing morphing wing architecture by testing both compliant and rigid-body mechanism-based morphing concepts and their related actuation, sensor, and control systems by monitoring aircraft weight and cost savings;
- investigating aircraft safety-related aspects by simulating system failures, such as jamming, runaways one engine loss, strong cross-wind, aeroelastic effects to validate fault tree analyses, and hazard assessments;
- including operational loads that apply hinge moment forces to the aircraft morphing surfaces, representative of the aerodynamics forces applied during the simulated flight test and driven by the flight simulation model;
- detailing cable routing and pathways;
- validating the electrical consumption of each actuation system, in stationary and dynamic conditions, and the required command to A/C surface in each test case.

2. Design of a morphing aileron

The design of a camber morphing aileron is following detailed as a reference case study for research into the subject. The aileron main functionalities such as roll maneuver are not modified. Conversely, with augmented capabilities integrated, the morphing aileron is deployed in cruise, through a symmetric deflection, to obtain a near optimum wing geometry enabling optimal aerodynamic performance. The design approach, including underlying concepts and analytical formulations, combines design methodologies and tools required to develop such an innovative control surface.

2.1. Multi-box structure design

Inner and medium wing regions where flap systems are generally located are growingly receiving considerable attention in research. That successful development was worth to be further investigated in order to understand its applicability to the whole wing span. It does then mean to verify the applicability of those concepts to the aileron region. This region plays a fundamental role for the aircraft roll control while is subjected to the external loads. Thus, during the preliminary design phase, it is important to consider some specific critical aspects: (i) The aileron constitutes a primary control surface, which is safety critical. Failure is a catastrophic event for the aircraft; (ii) the morphing capability is added to the conventional aileron which remains free to rotate around its main hinge axis; (iii) the aileron region constitutes a delicate zone from aero-elastic point of view; (iv) morphing will introduce normal modes driving flutter instability; (v) the wing tip region is characterized by very reduced space leading to a difficult integration of actuator and kinematic. This section details the design phases of the morphing aileron, spanning from preliminary numerical verifications to wind tunnel tests. The general morphing architecture and design process resemble the same philosophy developed for the SARISTU trailing edge. The device is aimed at working in cruise to modify a limited chord segment of the aileron, so to accomplish the aircraft weight variations following fuel consumption. However, during classical maneuver, this morphing part remains rigid and the aileron works in the usual manner. Such complex adaptive system has to meet specific requirements in terms of the aerodynamic target shape, stiffness distribution, and morphing controllability. In light of these considerations, an articulated mechanism was developed, in which each component have a predominant utility, but at the same time have to cooperate with the others in withstanding loads, distributing stress and driving the architecture in a controlled way from the baseline configuration to the target shapes (morphed down and morphed up). The proposed architecture was designed according to transport regional aircraft specifications. The morphing aileron is mainly composed of: (i) five segmented rib connected by means of rotational hinges positioned on the camber line creating a kinematic chain assuring enough structural robustness and transmitting deformation; (ii) spanwise stiffening elements such as spars and stringers in a multi-box arrangements; (iii) three servo-rotary actuators which drive the mechanism; (iv) a segmented skin ("armadillo-like" configuration) with silicon gap fillers to avoid discontinuities between adjacent parts and to ensure low friction sliding during morphing.

The geometrical external contour of the aileron constitutes the first step for its structural design. The rib mechanism uses therefore a three segment polygonal line to approximate the camber of the airfoil and to morph it into the desired configuration, while keeping approximately unchanged the airfoil thickness distribution. Each aileron articulated ribs (**Figure 11**) has been assumed to be segmented into three consecutive blocks (B0, B1, and B2) connected to each other by means of hinges displayed on the airfoil camber line (A and B) in a "finger-like" configuration. Moreover, non-consecutive rib plates are connected by mean of a link (L) that forces the camber line segments to rotate according to specific gear ratio and makes each rib equivalent to a single-DOF mechanism.

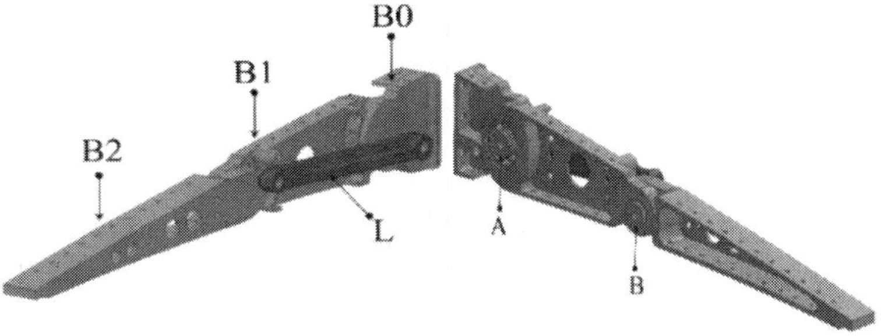

Figure 11. Morphing rib architecture: (a) blocks and links, (b) hinges.

The ribs' kinematic was transferred to the overall aileron structure by means of a multi-box arrangement (**Figure 12**). Each spanwise box of the structural arrangement is characterized by a single-cell configuration delimited along the span by homologue blocks of consecutive ribs, and along the chord by longitudinal stiffening elements (spars and/or stringers). Upon the actuation of the ribs, all the boxes are put in movement thus changing the external shape of the aileron; if the shape change of each rib is prevented by locking the actuation chain, the multi-box structure is elastically stable under the action of external aerodynamic loads. A four-bay (five-rib) layout was considered for an overall (true-scale) span of 1.5 meters. AL2024-T351 alloy was used for spars, stringers, and rib plates, while 17-4PH steel was used for ribs' links. Off-the-shelf airworthy components were properly selected for the bearing and bushings at the hinges and coupled to torsional springs to recover any potential free-play.

2.2. Actuation system design

The actuation system peculiarity resided in the fact that it is an un-shafted distributed servo-electromechanical arrangement deployed to achieve the aileron shape transition from the baseline configuration to a set of design target shapes in operative conditions moreover it is self-contained within the structure assuring a smooth surfaces exposed to the flow without fairing. The only kinematic mechanism that satisfies the target specifications is the oscillating glyph. The internal structure room defines the geometrical parameters which are directly related to the kinematic transmission ratio also defined as mechanical advantage (MA); furthermore, it is necessary to identify the number of actuators required to morph the aileron in particular due to small sizes near the tip, the last two bays could not be equipped with the kinematic. In **Figure 12**, it is shown that the first three ribs are drive by three individual actuators while the passive segment is slaved to the actuated one.

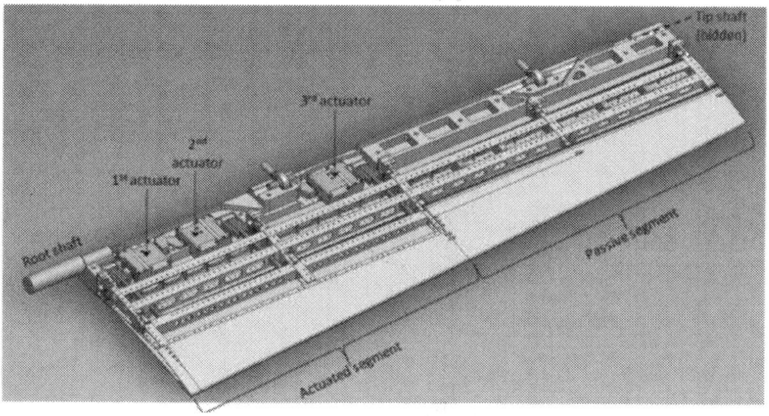

Figure 12. Internal view of the aileron with actuated and passive segment highlight.

A lightweight and compact leverage was investigated to activate the morphing aileron through EMAs. The deployment kinematics is based on a "direct-drive" actuation moving a beam rigidly connected to block B2 of **Figure 11**. The actuation beam transmits the actuation torque to the third segment of the rib, thus making it to rotate with respect to its original position. In particular, during morphing, the block B2 rotates around an instantaneous rotation centre. The instantaneous rotation center is here intended as the point in the moving plane around which all other points are rotating at a specific instant of time. As illustrated in **Figure 13(a)**, the trajectories of the points in the third block are all circles centered in this point. The determination of point V coordinates allows for the estimation of the actuation torque needed to withstand the aerodynamic loads acting on the morphing rib structure.

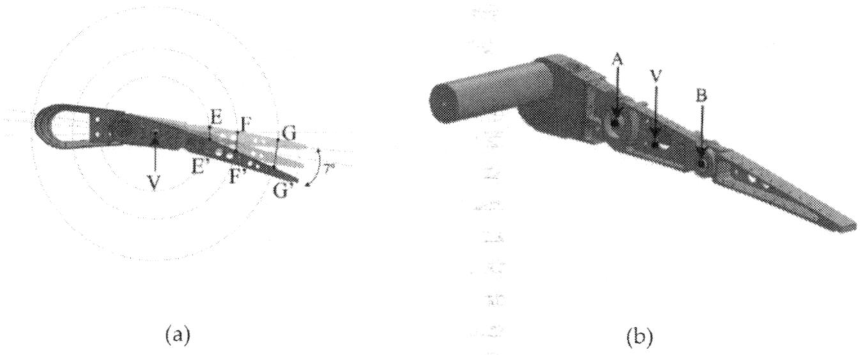

Figure 13. Circular trajectories of sample points (E, F, and G) during morphing (left) and position of hinges A, V, and B (right).

With reference to the **Figure 14**, the rotational motion of the actuation beam is provided by the crank rotation β which moves the carriage along its guide. A force F is thus generated by the contact between the carriage and the rail. By connecting the actuator shaft to the crank hinge O and the beam to the third rib segment (B2), the actuation torque is transmitted firstly to the crank and secondly to the rib rotating around the V in order to counterbalance the external moment $M_{rib\#3}$.

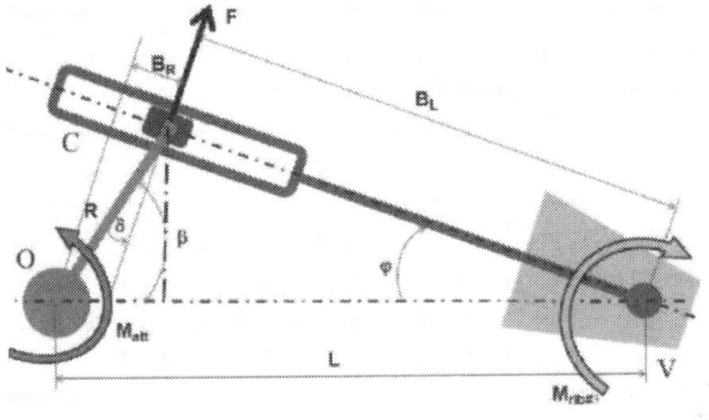

Figure 14. Oscillating glyph connected to the third rib segment of the morphing aileron [7].

The aileron shape can be, in this way, adaptively controlled to realize camber variations. The target morphing angles were derived as corresponding to a rigid rotation of a plain control surface comprised between -7° and +7°. The mechanical advantage of the mechanism (MA) can be written as follows:

$$MA = \frac{LOAD}{DRIVER} = \frac{M_{rib\#3}}{M_{att}} = \frac{F \cdot BL}{F \cdot BR} = \frac{BL}{BR} \tag{1}$$

where the $M_{rib\#3}$ is the external moment due to aerodynamic loads estimated with respect to the hinge V, while M_{att} is the actuation torque provided by the actuator in order to equilibrate the system. F is the force that the crank produces by means of the cursor, *BL* is the force arm, and *BR* is the crank projection along the guide. Equation (2) shows that the mechanical advantage only depends on the geometrical characteristics of the system. By combining geometrical terms, it follows:

$$\cot\varphi = \frac{L}{R \cdot \sin\beta} \cot\beta \tag{2}$$

This equation allows calculating the actuator shaft rotation (β) needed to achieve a given morphing angle (ϕ) of the rib block and hence of the entire mechanism. After estimating MA, it is possible to identify the actuation torque that actuator shall supply. Accordingly, the value of the force F shall be known in order to verify that the stress arising in the carriage moving into the rail, does not exceed design allowable. The actuation rod is then subjected to the simultaneous action of the force F and the external moment $M_{rib\#3}$, both producing bending stress. This indicates that actuation system design requires a trade-off between the mechanical advantage and the geometrical constraints limiting the actuator shaft rotation and L/R ratio. In order to mitigate the maximum counterbalancing load acting on the guide to equilibrate the aerodynamic moment, a fork-shaped crank coupled with a double sided linear guide was preferred, as shown in **Figure 15**.

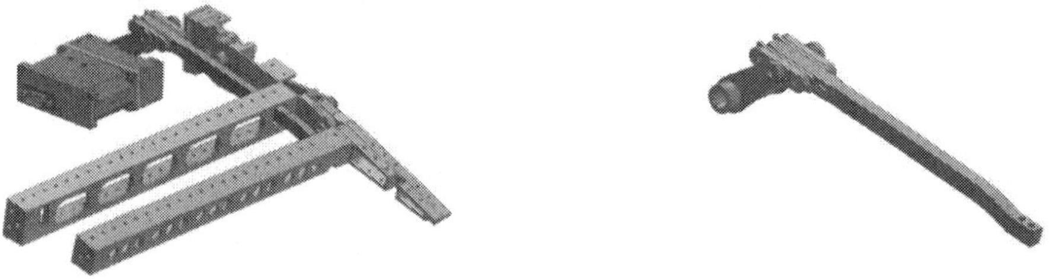

(a) (b)

Figure 15. Actuation system final architecture with high rigidity linear guide.

The VLM method was adopted to evaluate aerodynamic pressure distribution along the aileron in correspondence of each considered flight attitude (namely wing angle of attack, flight altitude, and speed) and aileron geometrical configuration. The obtained loads were considered for structural sizing and validation. A linear static analysis of the isolated actuation system mechanism by means of a FE simulation was, in a first approximation, performed. The aim of the numerical simulation was to verify if the static force acting on the linear guide was below the allowable value prescribed by the producer. In the real operative condition, the linear guide, being free to move, is not subjected to stress in the direction of motion. Force is transmitted in the vertical (with respect to the guide axis) and, partially, normal direction (with respect to the guide plane). For the current application, the actuator system was sized, referring to the jamming condition, considered as the most critical. In fact, as visible in **Figure 16**, the larger extent of the constraints (additional clamps) is expected to lead to higher stresses, locally (in the contact region) and distributed (overall). The actuation beam is then simultaneously loaded with the external aerodynamic moment, the vertical static force and a horizontal component (linked to the jamming), producing a pure bending with a higher stress level rather than the free guide. This effect was simulated by means of a perfect bonding between the rail and slider. The reaction force acting on the linear guide for a given aerodynamic moment was firstly evaluated and then compared to the expected actuation torque (**Figure 17**) multiplying by the crank length.

http://dx.doi.org/10.5772/63645

Figure 16. Stress contour on the linear guide element (max stress ~400 MPa).

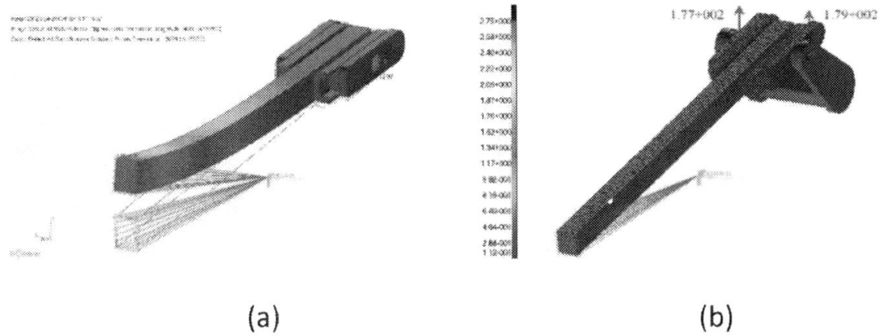

Figure 17. Beam displacement contour (a, left); guide reaction loads of 177 N and 179 N (b, right).

The finite element model of the entire aileron was then carried out. The FE model is representative of the three-dimensional drawings (CAD) of the entire aileron demonstrator. It includes main structural components such as segmented ribs and spars, actuation system leverage, and skin panels. Solid elements (CTETRA) were used for the mesh of the primary structure and the actuation leverage; meanwhile, beam elements (CBEAM) were used for modelling all the joints (fasteners, hinges, pins, and so on). FE model general data are recapped in **Table 1**.

FE model general data	
Number of elements	2.138 E+6
Number of nodes	1.393 E+6
Estimated DOFs	3.638 E+6
Total estimated volume (m^3)	6.785 E+6
Total estimated mass (kg)	21.00
Moment of inertia about aileron hinge-line (kg m^2)	0.403

Table 1. FE model characteristics.

The aileron primary structure is composed of ribs, actuation kinematic chains, spars, and skin. Aileron leading edge was not modelled for stress analysis purposes; however, it was considered only to properly evaluate the interface loads transmitted by the aileron to the wing box. In **Figure 18**, a global view of the aileron FE model is depicted, while in **Figure 19(a)** and **(b)**, details of rib and spars meshes are shown.

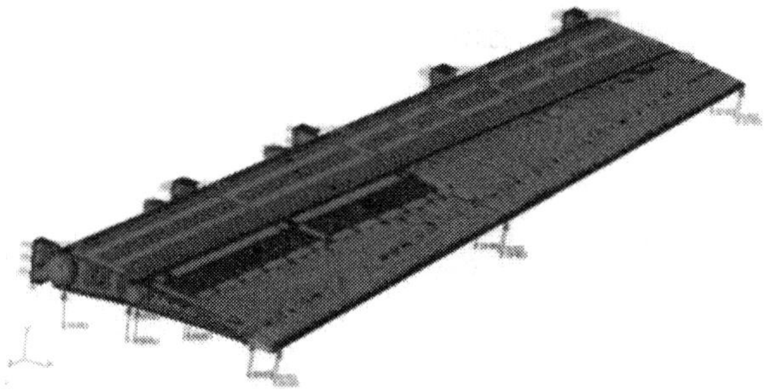

Figure 18. Aileron FE model.

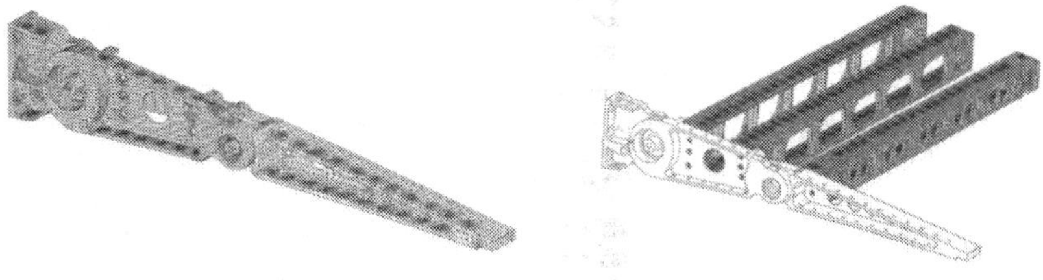

(a) (b)

Figure 19. (a) Aileron rib solid mesh (CTETRA), (b) spar solid mesh (CTETRA).

Main mechanical properties of the materials adopted for the aileron components are listed in the next table (**Table 2**).

Material (isotropic)	**E (GPa)**	**ρ (kg/m³)**	v	**Items**
Steel C50	220	7850	0.3	Beam of the actuation system, linear guide features, crank, and rib links
Al 2024-T351	70	2768	0.33	All other items

Table 2. Aileron components material.

All the components of the actuation system were connected to each other by means of several pins which were simulated using CBEAM elements (**Figure 20(a)** and **(b)**).

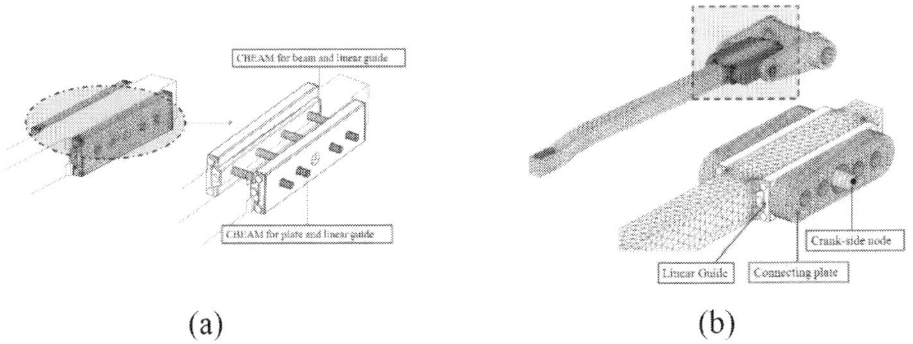

Figure 20. Connection pins between linear guides items (a) and detail of the local connection among the actuation kinematic parts (b).

Static analysis results have been here reported with reference to the limit load and ultimate load (1.5 times the limit load). In **Figure 21**, the global magnitude of the displacements exhibited by the aileron at limit load condition is shown. The maximum value (21.8 mm) is located at the trailing edge in proximity of the first bay.

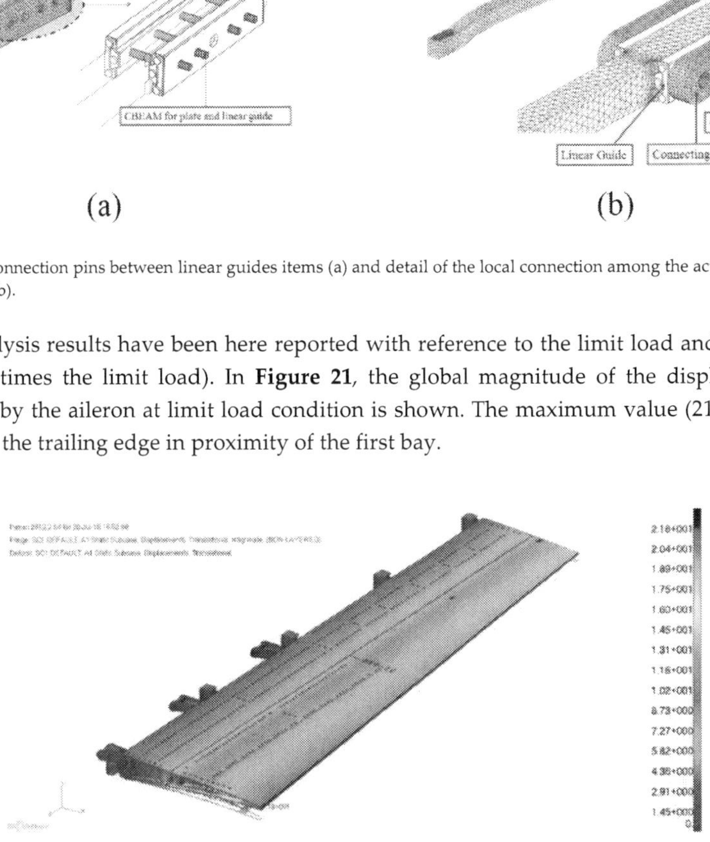

Figure 21. Global aileron displacement distribution at LL condition.

The stress distribution is characterized by concentrated peak around hinges and high solicitation of the actuation beam which is the most loaded components. Concerning the actuation levers, it is showed the typical stress distribution in bending; stress peaks greater than 350 MPa were found close to un-chamfered notches (**Figure 22(a)**). In addition, it is depicted (**Figure 22(b)**) the elements with stress level higher than 320 MPa. In this case, showing the most stressed elements is localized in a small area around the holes of the linkage between beam and spar and in proximity of the linear guides.

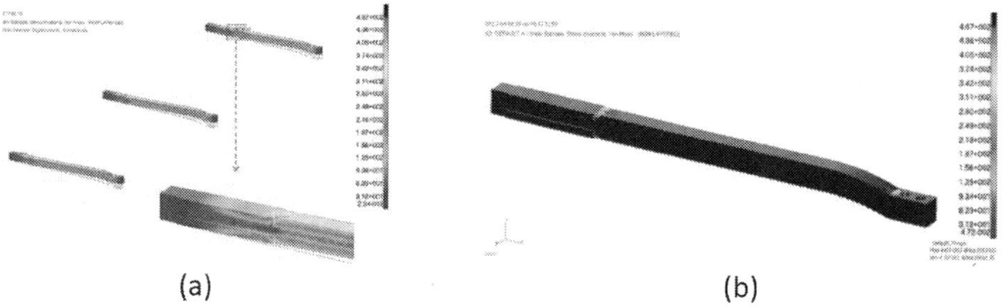

Figure 22. Global VM stress distribution on actuation beam at LL (a) and element stress distribution above threshold values of 320 MPa (b).

3. Prototyping and wind tunnel tests

On the basis of the numerical outcomes, the executive drawings of the prototype were produced and the aileron was then manufactured. Main structural parts are machined, while linear guides and actuators are components off-the-shelf (COTS). In the subsequent pictures, the segmented rib architecture, the actuation kinematic chain, and the final manufactured prototype (after painting) are shown. The morphing aileron was then integrated in a wing box and tested in wind tunnel at NRC (National Research Council of Ottawa, Canada), in the framework of the research program CRIAQ MDO505 involving Italian and Canadian university and research centre cooperation [8]. The aileron deflections are shown in **Figure 26**, and the integrated wing prototype is reported in **Figure 27**. The preliminary results obtained during wind tunnel tests were computed for baseline and morphed down configurations: lift versus angle of attack ($CL - \alpha$). (**Figure 28**); drag versus angle of attack ($CD - \alpha$) (**Figure 29**); drag polars ($CL - CL$) (**Figure 30**). The first one shows a typical linear trend. The curve slope (CL_{α}) remains unchanged and clearly by a morphing aileron deflection (from baseline to 6°), the camber increase (high α_{0L} and the curve moves in parallel upwards. The $CD - \alpha$ curve trend is reported in **Figure 29** for both unmorphed and morphed down configurations. The tendency shows that the minimum drag coefficient shift on the left as the morphing deflection increase leading to high CD_0. Finally, the drag polars are depicted in **Figure 30**. In this case, when a morphing deflection occur, the polar cross in correspondence of a pivot point for high CL while it moves on the right side of the Cartesian plane for low CL. This means that it is possible to identify an envelope curves which is the optimum one (dotted red line) (**Figures 23–25**).

http://dx.doi.org/10.5772/63645

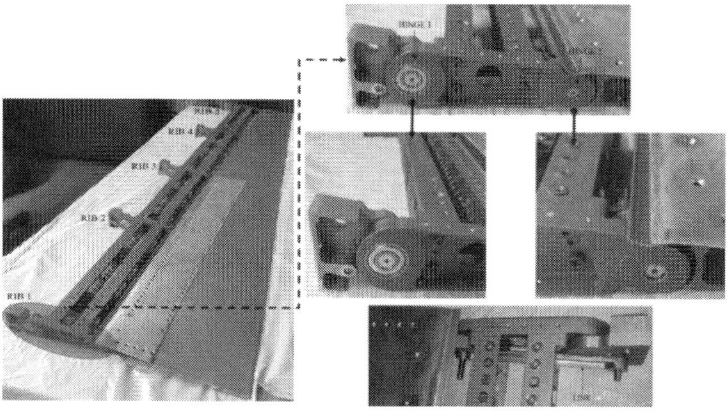

Figure 23. Aileron manufacturing with detail on hinges and rib.

Figure 24. Detail on aileron actuation system.

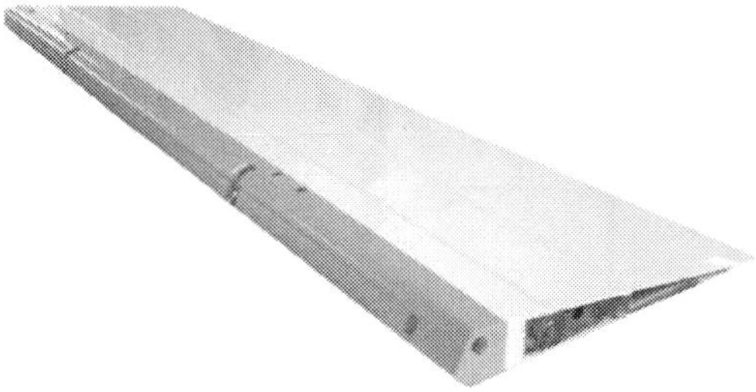

Figure 25. Photograph of the aileron prototype.

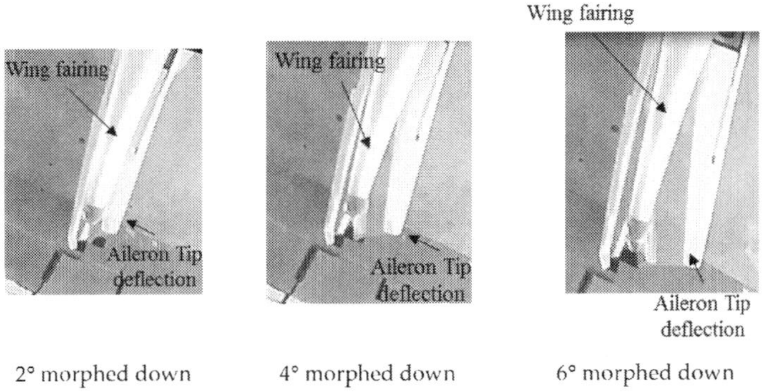

Figure 26. Morphing aileron at various deflections.

Figure 27. Complete CRIAQ wind tunnel test article including a morphing aileron [8].

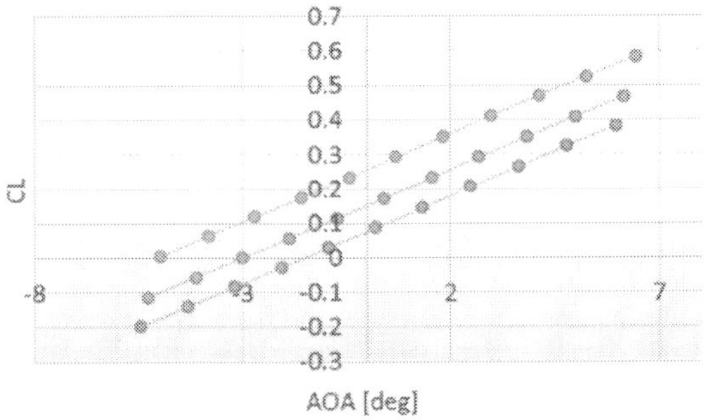

Figure 28. Lift coefficient versus angle of attack curve.

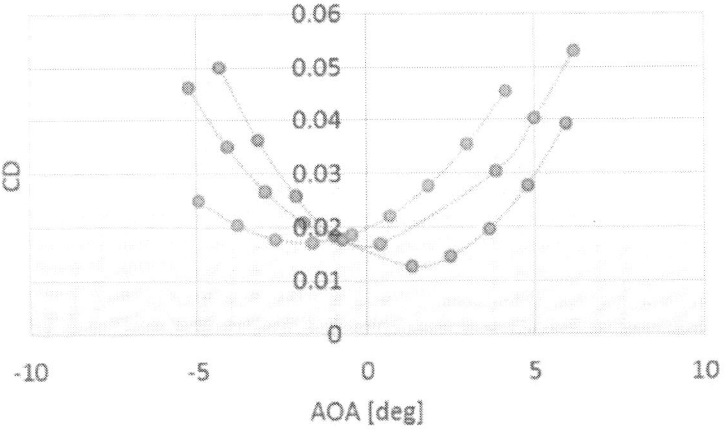

Figure 29. Drag coefficient versus angle of attack curve.

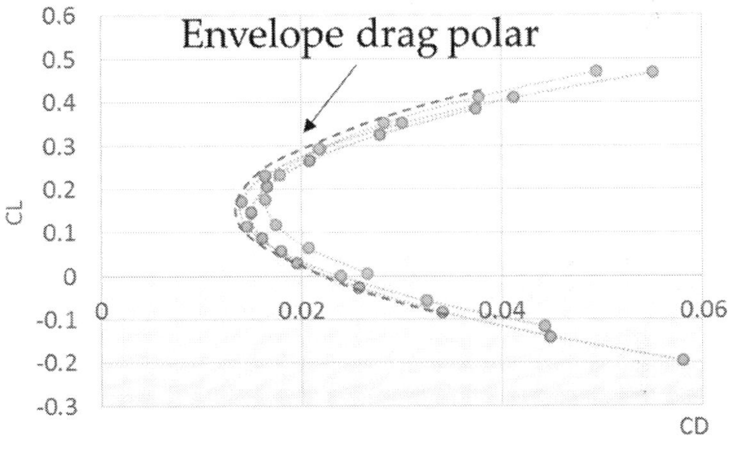

Figure 30. Drag polars with the envelope curve.

4. Conclusions

A self-contained morphing concept applied to a safety critical hinged control surface was presented in this chapter. In particular, a morphing aileron was investigated as an extension of an adaptive trailing edge in order to improve of L/D ratio and at the same time to preserve the conventional aileron functionality. The resulting morphed geometry, called "morphing aileron," ensures an augmented functionality with respect to a conventional "rigid" aileron. The device is able to rigidly rotate around main hinge axis and in addition will enable camber morphing. Being a safety critical surface, the structural design of a complete morphing aileron is rarely addressed in the literature. Such an original work provides thus evidence and arguments that contribute to the knowledge of morphing systems. Potentially suitable for static or dynamic purposes, the morphing aileron is an extension of the morphing trailing edge technology to the wing tip where small deflections could bring significant aerodynamic benefits. It has been designed for a symmetrical deflection during cruise in order to compensate A/C weight variation due to fuel burned. In such a manner, it is aimed to increase aerodynamic efficiency (reduce drag) in off design points. Additionally, the deflection of a morphing aileron it is expected to redistribute the spanwise wing distribution in order to reduce wing root bending moment. On the other hand, by increasing actuator bandwidth, it can be tailored to reduce peak stress from gust.

In order to deflect a "finger-like" rib architecture, a compact electromechanical actuation based on double-sided guides and a fork-shaped crank has been designed. Advanced finite element model in order to validate the structure at limit and ultimate loads have been carried out setting all the details necessary to produce a laboratory demonstrator. This one was assembled and tested, proving the effective functionality of the concept. Finally, wind tunnel tests assessing the aerodynamic trend of such innovative architectures have been reported. The idea herein described leads the way to further researches aimed at enhancing the TRL of the concept. To this aim, some remarks should be done on the most critical aspects of the current device. In particular, future steps may be: (i) an embedded sensing network for enhanced control in order to assure the achievement of the target aero-shapes; (ii) actual shapes evaluation and comparison with expected aero-shapes; (iii) aerodynamic benefits comparison between rigid and morphing aileron; (iv) morphing aileron-related (wing and A/C) performance benefits estimations; (v) enhanced design with topology optimization; (vi) segmented skin aerodynamics comparison with a tailored complaint skin technology; (vii) high-speed simulations and tests.

Author details

Ignazio Dimino[1*], Gianluca Amendola[1], Francesco Amoroso[2], Rosario Pecora[2] and Antonio Concilio[1]

*Address all correspondence to: i.dimino@cira.it

1 CIRA, The Italian Aerospace Research Centre, Adaptive Structures Division, Capua, Italy

2 University of Napoli, "Federico II", Industrial Engineering Dept, Aerospace Division, Napoli, Italy

References

[1] Valasek, J., *"Morphing Aerospace Vehicles and Structures,"* John Wiley & Sons, Ltd., United States (2012)

[2] Wölcken, P.C., Papadopoulos, M., *"Smart Intelligent Aircraft Structures (SARISTU)"*, Proceedings of the Final Project Conference, Springer, Germany (2015). ISBN: 978-3-319-22413-8.

[3] Recksiek, M., *"Advanced High Lift System Architecture with Distributed Electrical Flap Actuation,"* AST 2009, March 29–30, Hamburg, Germany.

[4] Dreßler, U., Take-off and landing configurations, DaimlerChrysler Aerospace, March 1999.

[5] Rudolph, P.K.C., *"High-Lift System on Commercial subsonic Airlines,"* NASA Report 4746, September 1996.

[6] Derrien, J.C., *"Electromechanical Actuator (EMA) Advanced Technologies for Flight Controls,"* Presented at 28th International Congress of the Aeronautical Sciences (ICAS 2012).

[7] Dimino, I., Flauto, D., Diodati, G., Concilio, A., Pecora, R., *"Actuation System Design for a Morphing Wing Trailing Edge,"* Recent Patents on Mechanical Engineering, Volume 7, 2014, pp. 138–148.

[8] Kammegne, M.J.T., Botez, M.R., Mamou, M., Mebarki, Y., Koreanschi, A., Gabor, O.S., Grigorie, T.L., *"Experimental Wind Tunnel Testing of a New Multidisciplinary Morphing Wing Model"*, Proceedings of the 18th International Conference On Mathematical Methods, Computational Techniques and Intelligent Systems (MAMECTIS 2016).

Chapter 6

Self-Healing Control Framework Against Actuator Fault of Single-Rotor Unmanned Helicopters

Xin Qi, Zhong Liu, Yuqing He, Liying Yang,
Yuqing He and Jianda Han

Additional information is available at the end of the chapter

http://dx.doi.org/10.5772/62480

Abstract

Unmanned helicopters (UHs) develop quickly because of their ability to hover and low speed flight. Facing different work conditions, UHs require the ability to safely operate under both external environment constraints, such as obstacles, and their own dynamic limits, especially after faults occurrence. To guarantee the postfault UH system safety and maximum ability, a self-healing control (SHC) framework is presented in this chapter which is composed of fault detection and diagnosis (FDD), fault-tolerant control (FTC), trajectory (re-)planning, and evaluation strategy. More specifically, actuator faults and saturation constraints are considered at the same time. Because of the existence of actuator constraints, usable actuator efficiency would be reduced after actuator fault occurrence. Thus, the performance of the postfault UH system should be evaluated to judge whether the original trajectory and reference is reachable, and the SHC would plan a new trajectory to guarantee the safety of the postfault system under environment constraints. At last, the effectiveness of proposed SHC framework is illustrated by numerical simulations.

Keywords: fault detection and diagnosis, fault-tolerant control, invariant set, self-healing control framework, trajectory (re-)planning

1. Introduction

Safe operation and reliability are important performances of unmanned aerial vehicles (UAVs). Unmanned helicopters (UHs) as a kind of UAVs develop quickly because of their ability to hover and low speed flight. Nevertheless, traditional control and planning strategies cannot guarantee their safety and reliability in the face of malfunctions such as sensor faults, actuator faults, and component faults. Furthermore, taking physical limits such as actuator constraints and state constraints into account, the situation would be more serious. In order to ensure the reliability and safety of UHs, fault detection and diagnosis (FDD) and fault-tolerant control (FTC) approaches become focus of research and many related research results have been presented [1]. However, physical limits are not well considered in conventional FTC approaches which may affect their efficiency. On the other hand, besides UH's dynamic limits, external environment constraints, such as obstacles, also affect UH's safety. However, traditional planning methods cannot consider environment constraints and vehicles' limits at the same time which makes these methods helpless in the face of faults. In this chapter, a self-healing control (SHC) framework is proposed against actuator faults and constraints of single-rotor UHs under external environment constraints. The SHC framework is shown in **Figure 1** which involves FDD module, reconfigurable controller, trajectory (re-)planning module, and evaluator module.

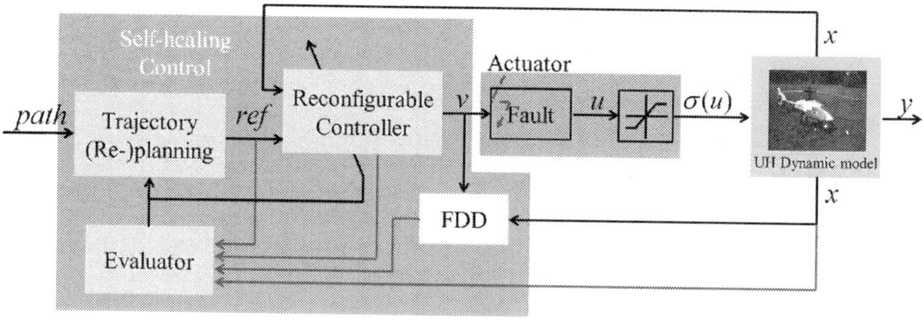

Figure 1. SHC framework.

The tasks of the above modules are as follows:

1. FDD module: Estimate the actuator healthy coefficient (AHC) [2] in real time. AHCs can indicate healthy conditions of actuators and actuator fault information is the basis of controller reconfiguration.

2. Reconfigurable controller: This module realizes the same functions as conventional FTC approaches and considers actuator constraints at the same time. After fault occurrence, the fault-free controller will be configured as a postfault controller according to AHCs provided by FDD module to guarantee the stability of the postfault UH system.

3. Trajectory (re-)planning module: Compute realizable trajectory based on desired path and UH dynamic model under environment constraints. This module calculates a feasible trajectory under both fault-free and postfault conditions. Furthermore, controller references are computed directly in this module which can be used by reconfigurable controller.

http://dx.doi.org/10.5772/62480

4. Evaluator module: Evaluate system performance of UHs according to AHCs, UH dynamic model, and controller and trajectory information. If the original controller and trajectory are not feasible, this module will ask related modules to reconfigure controller and replan trajectory.

The remaining part of this chapter is organized as follows: single-rotor UH and AHC models are simply introduced in Section 2; Section 3 investigates FDD approach against actuator faults based on extended Kalman filter (EKF) and linear neural network (LNN). In Section 4, the postfault controller is designed to guarantee the stability of UH system under both actuator faults and constraints; Section 5 presents invariant-set based planning (ISBP) approach that can compute controller reference according to desired path and UH dynamic model. Evaluation strategy is introduced in Section 6 and numerical simulations are shown in Section 7. Finally, Section 8 summarizes the chapter with conclusions.

2. Model of UH and AHCs

2.1. UH model

The 6-DOF dynamics of UH is given by the following Newton–Euler equations:

$$\dot{V}_b = -\omega_b \times V_b + \frac{F_b}{m} + \frac{F_g}{m}$$
$$\dot{\omega}_b = -[M_b - \omega_b \times (I_b \omega_b)] / I_b$$

where $V_b=[u\ v\ w]$ is the velocity vector, $\omega_b=[p\ q\ r]$ is the angular velocity vector, F_b is the aerodynamic force vector, F_g is the gravity force vector, I_b is the moment of inertia matrix, M_b is the aerodynamic moment vector, and m is the UH mass.

Aerodynamic forces and moments can be calculated by

$$F_b = f_f(\delta_{\text{lon}}, \delta_{\text{lat}}, \delta_{\text{col}}, \delta_{\text{ped}}, \omega_r)$$
$$M_b = f_m(\delta_{\text{lon}}, \delta_{\text{lat}}, \delta_{\text{col}}, \delta_{\text{ped}}, \omega_r)$$

where δ_{lon}, δ_{lat}, δ_{col}, and δ_{ped} are the longitude, latitude, main rotor collective pitch, and tail rotor collective pitch control inputs, respectively. ω_r is control input of main rotor rotating speed. Typically, f_f and f_m are nonlinear functions and their details can be found in [3, 4].

Note that δ_{lon}, δ_{lat}, δ_{col} are the nominal control inputs and they do not represent the actual actuator outputs. The relationship between the nominal and actual inputs is determined by the structure of swashplate [5] and their values can be converted to each other.

In order to guarantee the accuracy of AHC estimation, nonlinear model is used to design FDD approach. However, the above nonlinear model is too complex to design EKF filter so that a simplified nonlinear discrete-time model is used [6]:

$$\begin{cases} x(k) = f(x(k-1), u(k-1)) + \omega(k-1) \\ y(k) = h(x(k)) + v(k) \end{cases} \tag{1}$$

where $x \in R^n$ is the system state vector, $y \in R^p$ is the system output vector, $u \in R^m$ is the system control input vector, ω and v are the process noise and measurement noise that are assumed to be Gaussian white noise with zero mean and uncorrelated from each other. Considering swashplate configuration [5], the system control input vector is $u=[\theta_l, \theta_r, \theta_b, \theta_t, \omega_r]$ where θ_l, θ_r, θ_b are left, right, back swashplate actuator outputs, and $\theta_t=\delta_{ped}$ is the tail actuator output.

On the other hand, the linear dynamic model is used to design controller and trajectory planning which is shown as follows:

$$\begin{cases} x_h(k+1) = A_h x_h(k) + B_h u_h(k) \\ y_h(k) = C_h x_h(k) \end{cases} \tag{2}$$

where $x_h \in R^n = x - x_{\text{trim}}$, $y_h \in R^p = y - y_{\text{trim}}$, $u_h \in R^m = u - u_{\text{trim}}$, and x_{trim}, y_{trim}, u_{trim} are the trim values. A_h, B_h, and C_h are all constant matrices.

2.2. AHC model

Suppose v is the control signal given by controller called control variable and u is the actual actuator output called manipulated variable, their relationship is represented by

$$u(k) = \Lambda v(k) + (I - \Lambda)\bar{u}$$

where Λ is a diagonal matrix with $\Lambda = \text{diag}(\lambda_1, \ldots, \lambda_m)$, $\lambda_i \in [0, 1]$, and $\bar{u} \in R^m$ is a constant vector. In addition, λ_i and $\bar{u}_i$ are the proportional effectiveness and fault bias of the ith actuator's AHC. $\lambda_i=1$ and $\bar{u}_i=0$ represent that the ith actuator is fault-free; otherwise, the ith actuator has fault. Thus, the fault condition of actuators can be represented by AHCs.

3. EKF- and LNN-based FDD approach

Because of the serious nonlinear coupling between all control channels, we assume that an actuator has fault whereas others are still well, and the actuator for rotate speed control, ω_r, is always well. Here, the multiplicative fault of an actuator as a parameter that is indicated by the AHC for this kind of fault is more likely to happen. The following will introduce the realization of fault diagnosis in the order of fault identification module, fault isolation module, and fault detection module. In this section, $x(k)$ is written as x_k for simplicity and $\hat{x}_{k|k-1}$ represents the pre-estimation value of x_k based on the information of $k-1$ step.

3.1. Fault identification module

The task of fault identification is to determine the size of the fault and the fault time after fault isolation. The size of a fault is indicated by the proportional effectiveness of AHC, so we should estimate the actual manipulated variable vector u according to state estimations from EKF and actual control signal v to calculate and obtain the proportional effectiveness.

As shown in Eq. (1), in a typical nonlinear discrete-time system, the input vector is obtained from actuators, and the current state vector x_k is calculated based on the state vector of the last step. The estimations of $\hat{x}_{k|k}$ and $\hat{x}_{k-1|k-1}$ can be obtained by EKF [7]. Therefore, the state equation can be transformed into the equation as follows:

$$\hat{x}_{k|k} = M(\hat{x}_{k-1|k-1})u + N(\hat{x}_{k-1|k-1}) \tag{3}$$

For the calculation, processes of forces and torques of the simplified nonlinear model have been linearized, Eq. (3) is a linear equation set, and the values of M and N are connected with state estimations of the last step only.

Typically, Eq. (3) is an overdetermined linear equation set and could not be solved directly [8]. After knowing which actuator is faulty, many output estimations of this actuator can be obtained by the faultless manipulated variables of other actuators according to Eq. (3). Multiply the sample column vectors formed by these estimations by a weight matrix and correct this result with an offset. Then, the manipulated variable estimation of this faulty actuator can be obtained.

To obtain the above weight matrices and offsets, consider a linear neuron as a simple LNN, and its input–output relation is expressed as follows:

$$a = f(Wp + b) = Wp + b$$

where a is the output vector, p is the input sample vector, W is the weight matrix, and b is the offset. The weight matrices and offsets can be obtained by training. Assume that an actuator has a fault, and let the output estimations of this actuator be a set of input of the linear neuron. After simulation, multiple sets of input samples can be obtained, and let the actual manipulated variable be the target value to train the weight matrix and offset of this actuator. Determine the weight matrix and offset of every faulty actuator, and make use of them in the online experience to calculate the estimation of every manipulated variable. The above process is closely related to the estimation process of EKF; more certainly, EKF and LNN complete the joint estimation of system states and AHCs together.

3.2. Fault isolation module

The task of fault isolation is to determine the location and type of the fault after fault detection. The structure of this module is shown in **Figure 2** to record the completion time and pass the serial number of the faulty actuator to fault identification module.

Four fault identification modules are in parallel in this module, and we assume that there are different actuator faults in different identification modules. Therefore, every fault identification module can obtain the proportional effectiveness of AHC called isolation AHC. Make use of the isolation AHC of the ith actuator and let these values of other actuators be both 1 to calculate the manipulated variables called isolation-manipulated variables according to control variable vector v. Based on these variables, we can obtain the system state vector and define it as x_{iso}. In this way, we define the isolation residual of the ith actuator as follows:

$$\text{res}_i = (x_{\text{iso}} - x_{\text{est}})^T (x_{\text{iso}} - x_{\text{est}})$$

where x_{est} is state estimation. If the isolation residual of an actuator is less than the isolation residuals of others, there will be fault in this actuator. For example, if res_i is less than other isolation residuals, the ith actuator can be considered to be faulty.

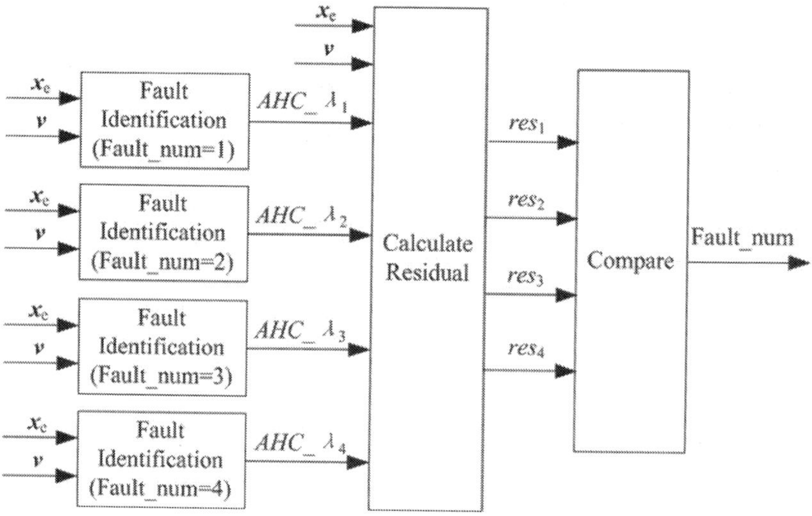

Figure 2. The structure of fault isolation.

3.3. Fault detection module

The basis of the above two modules is that the fault is detected quickly and accurately. Known from the EKF process, the filter obtains the state vector pre-estimation $\hat{x}_{k|k-1}$ by the estimations of the last step and updates $\hat{x}_{k|k-1}$ by the measurement information to obtain the current estimation of the state vector $\hat{x}_{k|k}$. If there is a fault in the actuator, there will be difference between the pre-estimations of states and the actual states with fault. However, the estimations of states would track the actual states after the update. Define a filter residual as follows:

$$r_k = \hat{x}_{k|k} - \hat{x}_{k|k-1}$$

Perform the weighted sum of squared residuals (WSSR) operation [9]:

$$\mathrm{WSSR}_k = r_k^T \Sigma r_k$$

where the weight matrix $\Sigma = \mathrm{diag}(\sigma_1^2, \ldots, \sigma_n^2)$ and σ_i is the standard deviation of the filter residual of the ith state variable to make the residual that can indicate faults work.

According to the threshold value γ set in advance, the criteria for fault detection are as follows: if $\mathrm{WSSR}_k > \gamma$, fault; if $\mathrm{WSSR}_k \le \gamma$, no fault.

4. Reconfigurable controller design

In this section, AHC-based antiwindup controller design method is introduced which can be achieved by solving a set of linear matrix inequalities (LMIs). At the same time, the related invariant-set-based safety region is also calculated. The definition of invariant set and the controller design under actuator constraints are given as follows.

4.1. Invariant set

Considering a linear system

$$x(k+1) = Ax(k)$$

the (positively) invariant set can be given by the following definition.

Definition 1 [10]: The set $\mathrm{S} \subset R^n$ is said invariant for the above system if for all $x(0) \in \mathrm{S}$ the solution $x(k) \in \mathrm{S}$. If $x(0) \in \mathrm{S}$ implies $x(k) \in \mathrm{S}$ for $k > 0$, then we say that S is positively invariant.

Based on Definition 1, the definition of robustly controlled invariant set for the following system is provided:

$$\begin{cases} x(k+1) = Ax(k) + Bu(k) + D\omega(k) \\ y(k) = Cx(k) \end{cases}$$

Definition 2 [10]: The set $\mathsf{S} \subset R^n$ is said robustly controlled invariant for the above system if there exists a continuous feedback control law $u(k)=Kx(k)$, which ensures the existence and uniqueness of the solution on R^+ such that S is positively invariant for the closed loop system.

An invariant set S can be constructed by the following theorem.

Theorem 1 [11]: The set $\mathsf{S}_\rho = \{x(k) \in R^n \mid V(x(k)) \leq \rho\}$ with $\Delta V(x(k)) \leq 0$ for all $x(k) \in \mathsf{S}_\rho$ is positively invariant where V is a function.

For convenience, ρ is chosen equal to 1 and simplify S_ρ as $\mathsf{S} = \{V(x(k)) \leq 1\}$. Hence, the construction of S finds a function V, which implies $\Delta V(x(k)) \leq 0$ when $V(x(k)) \leq 1$. Obviously, if we assume $V(x(k)) \geq 0$, V would be a candidate of the Lyapunov function. More specially, if the Lyapunov candidate function is in quadratic form such that $V(x(k)) = x^T(k)Px(k)$, the invariant set will be an ellipsoid and this nature will be used later.

4.2. Controller design under actuator constraints

Consider linear discrete UH model, Eq. (2), with actuator constraints:

$$\begin{cases} x_h(k+1) = A_h x_h(k) + B_h \sigma(u_h(k)) \\ y_h(k) = C_h x_h(k) \end{cases}$$

Where $\sigma(u_h(k))$ represents constrained control inputs such that $-1 \leq u(k) \leq 1$ and the saturation feature can be defined by $\sigma(u_i(k)) = \mathrm{sgn}(u_i(k))\min\{1, \ |u_i(k)|\}$, where $\mathrm{sgn}(\cdot)$ is a sign function and $i=1, \ldots, m$. Then, considering AHC model, the above system can be rewritten as

$$\begin{cases} x_h(k+1) = A_h x_h(k) + B_{h0}\sigma(v(k)) + B_{hf}\bar{u} \\ y_h(k) = C_h x_h(k) \end{cases}$$

Where $B_{h0} = B_h \Lambda$ and $B_{hf} = B_h(I - \Lambda)$. In the following discussion, assume the number of $\lambda_i = 0$ is m_f, which implies m_f actuators cannot respond to the control signal and $m_0 = m - m_f$ is the number of fault-free/partial-fault actuators that can respond to the control signal. In other words, m_0 represents the number of nonzero columns of matrix B_{h0}. In this way, the postfault system with different actuator faults can be represented.

In order to achieve set-point tracking offset free and compensate actuator saturation, an integrator with a saturation compensator is introduced as follows:

$$e(k+1) = e(k) + T_s\left(\text{ref} - C_h x_h(k)\right) - T_s E_c \varphi(v(k))$$

where $e(k)$ is the integrator state vector, E_c is the pending compensator matrix, $\varphi(v(k))=v(k)-\sigma(v(k))$ is the difference of controller outputs and actuator outputs under saturation [12], $\text{ref} \in \Omega_{\text{ref}} = \{\text{ref} \mid \text{ref}^T Q_r \text{ref} \leq 1\}$ is a set-point reference, Q_r is a constant matrix, and T_s is sampling period. Thus, a new open-loop system with extend state vector $x(k)=[x_h^T(k) \;\; e^T(k)]^T$ can be expressed by

$$\begin{cases} x(k+1) = A_e x(k) + Bv(k) + D\omega(k) - (B + RE_c)\varphi(v(k)) \\ y(k) = Cx(k) \end{cases} \tag{4}$$

where $\omega(k)=[\bar{u}^T \;\; \text{ref}^T]^T$,

$$A_e = \begin{bmatrix} A_h & 0 \\ -T_s C_h & I \end{bmatrix}, B = \begin{bmatrix} B_{h0} \\ 0 \end{bmatrix}, D = [D_1 \; D_2] = \begin{bmatrix} B_{hf} & 0 \\ 0 & T_s I \end{bmatrix},$$
$$R = \begin{bmatrix} 0 \\ T_s I \end{bmatrix} \text{and}\, C = [C_h \quad 0]$$

Note that, if $m_f = 0$, $\omega(k)=\text{ref}$, then $D=[0\,(T_s I)^T]^T$ will hold.

Based on linear quadratic regulator (LQR) theory, state-feedback controller $v(k)=Kx(k)$ can be calculated. Thus, the closed-loop postfault system can be expressed by

$$\begin{cases} x(k+1) = Ax(k) + D\omega(k) - (B + RE_c)\varphi(v(k)) \\ y(k) = Cx(k) \end{cases}$$

where $A=(A_e + BK)$. In order to guarantee the closed-loop system stability and to track offset-free under external input $\omega(k)$ and actuator constraints, the following theorem is proposed. As a basis, first a lemma is given which defines a set of system states related to actuator saturation.

Lemma 1 ([12]): Define the following polyhedral set with matrix $G \in R^{m_0 \times (n+p)}$:

$$\mathcal{E} = \{x(k) \mid |(K_i - G_i)x(k)| \leq 1, i = 1, \ldots, m\}$$

where K_i and G_i represent the ith row of matrices K and G. If $x(k) \in \mathcal{E}$, then the relation

$$\varphi(Kx(k))^T T\left[\varphi(Kx(k)) - Gx(k)\right] \leq 0$$

is verified for any positive-definite matrix $T \in R^{m_0 \times m_0}$.

Based on this lemma, saturation compensator can be designed in the following theorem.

Theorem 2: Given LQR parameter matrices Q_u, Q_x, if there exist a symmetric positive-definite matrix $W \in R^{(n+p)\times(n+p)}$, matrices $X \in R^{m_0 \times (n+p)}$, $Y \in R^{m_0 \times (n+p)}$, $Z \in R^{p \times m_0}$, diagonal positive-definite matrices $S \in R^{m_0 \times m_0}$, $W_r \in R^{p \times p}$, and a positive scale η satisfying

$$\inf_{W,X,Y,Z,S,W_r} \eta$$

$$\begin{bmatrix} 1 & X_i - Y_i \\ * & W \end{bmatrix} \geq 0, I = 1,\ldots,m_0$$

$$\begin{bmatrix} -W & Y^T & 0 & 0 & WA^T & -C^T & X^T & W \\ * & -2S & 0 & 0 & -SB^T - Z^T R^T & 0 & 0 & 0 \\ * & * & -\eta & 0 & D_1^T & 0 & 0 & 0 \\ * & * & * & -W_r & D_2^T & I & 0 & 0 \\ * & * & * & * & -W & 0 & 0 & 0 \\ * & * & * & * & * & -I & 0 & 0 \\ * & * & * & * & * & * & -Q_u^{-1} & 0 \\ * & * & * & * & * & * & * & -Q_x^{-1} \end{bmatrix} < 0$$

where X_i, Y_i represent the ith row of matrices X, Y and $K = XP$, $G = YP$. Then, the saturation compensator will be $E_c = ZS^{-1}$ and the closed-loop postfault system will be stable if $x(k)$ inside the stability domain $\mathsf{S} = \{x(k) \mid x^T(k)Px(k) \leq 1\}$ with $P = W^{-1}$. Note that S is defined as the safety region in this chapter. Ω_{ref} is also achieved with $Q_r = W_r^{-1}$. The proof can be found in [13].

According to Theorem 2, the safety region S is an invariant set which means that if initial states and steady states of a system are inside the set, the state trajectory from initial states to steady states will also be inside the set. In this way, a postfault system with a fault-tolerant controller can be seen as a new closed-loop system with the initial state $x(k_f)$, where k_f is instant when the fault is detected Clearly, a postfault system will be safe if initial state $x(k_f)$ is inside the safety region S.

On the other hand, the related admissible set of reference Ω_{ref} is also achieved by Theorem 2 and it has a closed relationship with the controller. In the next section, the controller design and the reference design will be composed together.

http://dx.doi.org/10.5772/62480

5. ISBP approach

This section proposes an ISBP approach that can plan a feasible trajectory based on the desired path nodes under both external environment constraints and their own dynamic limits [14]. In order to consider environment constraints and dynamic constraints at the same time, invariant set is used as a bridge to connect the two kinds of constraints because the invariant set is calculated based on the Lyapunov function that is linked to dynamic model, and on the other hand, the invariant set has certain geometrical shapes, such as ellipsoid, so that it can easily be represented in work space with obstacles.

Note that, in this section, we have assumed that the heading of UH is kept to be 0, pitch angle and roll angle of UH are kept small so that the position of UH in the world coordinate system can be considered as the integration of velocities in the body coordinate system.

5.1. Reachable set

Considering System (4) with controller $v(k) = Kx(k)$, the definition of reachable set is given.

Definition 3 [15]: The reachable set S_r is defined as

$$S_r = \{ y(k) = Cx(k) \mid x(k) \in S_{rx} \}$$

$$S_{rx} = \left\{ x(k) \middle| \begin{array}{c} x(k), \omega(k), \text{ref satisfy}(4), \\ v(k) = Kx(k) \in \Omega_v, \\ \text{ref} \in \Omega_{\text{ref}}, x(0) = 0, k \geq 0 \end{array} \right\}$$

where $\Omega_v = \{v(k) \mid -1 \leq v_i \leq 1, i = 1, \ldots, m_0\}$.

5.2. ISBP method

For the convenience of discussion, a 2D work space is used in the following whereas the proposed ISBP approach can be expanded in 3D condition directly. Consider a preknown environment as shown in **Figure 3**, where the initial location is L_0 and the goal location is L_5. Based on the preknown environment, a lot of path planning approaches can be used to find a feasible path against external environment constraints such as obstacles. Suppose this path is represented by a series of nodes such as $L_0 \sim L_5$, as shown in **Figure 3**.

The target of the proposed ISBP approach is to calculate feasible controller reference inputs to move the UH from the initial location to the goal location based on the feasible path and satisfy both external environment constraints and UH's limits.

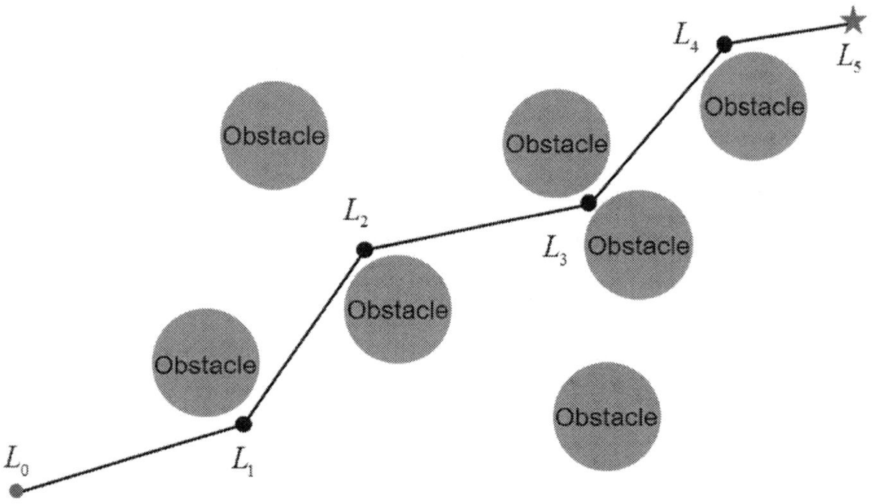

Figure 3. Two-dimensional environment and feasible path.

Consider a simple example in the preknown environment as shown in **Figure 4(a)** where the initial location is L_0 and the goal location is L_1. Suppose a feasible path is represented by two nodes L_0 and L_1 and their connection is shown in the same figure.

Clearly, the work space is state space; thus, the location in work space is equality to the state such as the initial state, which is equal to L0. Thus, an invariant set S can be constructed according to plant dynamic model (4) under controller $v(k)=Kx(k)$, which also guarantees obstacle avoidance. Suppose the invariant set is S_0 and the related reachable set is S_{r0} as shown in **Figure 4(b)**. Clearly, because of the existence of constraints such as obstacles and UH's limits, the invariant set S_0 may not contain the goal location L_1 as shown in **Figure 4(b)**. Closely, the goal location L_1 is also outside the reachable set S_{r0}, which means L_1 is unreachable under this condition. Thus, the system states should be continued moving.

Suppose the intersection of reachable set S_{r0} and feasible path $\overline{L_0L_1}$ is c_1 as shown in **Figure 4(b)**. In order to move state from c_1 to L_1, a new invariant set, whose center is c_1, is required. Suppose the state of the new system is $\bar{x}(k)=x(k)-x_{c1}$, where x_{c1} is a constant vector whose position is equal to c_1 and other elements are 0. Based on the new system, we can construct the second invariant set S_1 whose center is c_1 as shown in **Figure 4(c)**. The related reachable set S_{r1} is also shown. Obviously, the goal location L_1 is inside the reachable set S_{r1} now, which means L_1 is reachable under this condition. In this way, we can choose the center of S_1, c_1, and the goal location L_1 as a sequence of the controller reference input. Finally, Figure 4(d) shows the practical trajectory that is obtained by the UH actual operation.

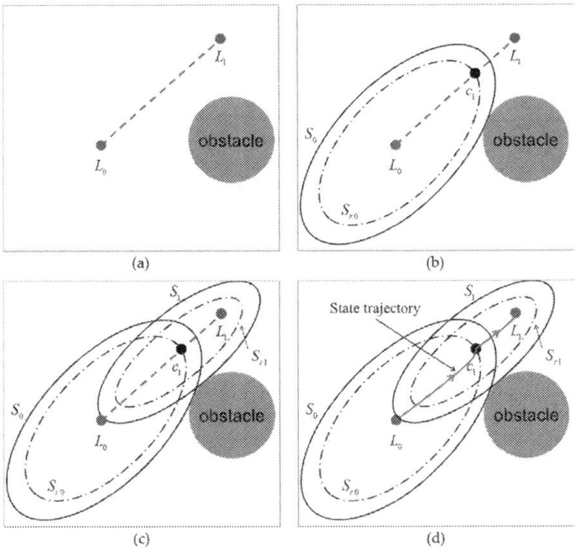

Figure 4. ISBP in 2D condition.

5.3. Calculation of the invariant set and the reachable set

According to Theorem 2, the invariant set S and the reference set Ω_{ref} are obtained. Assume the reachable set S_r defined by Definition 3 is included in the invariant set S such that $S_r \subset S$. Because the reference is bounded such as ref$\in \Omega_{\text{ref}}$ and the integrator can guarantee all references to be reachable such as $y(\infty)=\text{ref}$, the reachable set S_r can be denoted by $S_r=\{\text{ref} \mid \text{ref}^T Q_r \text{ref} \le 1\}$. Thus, the invariant set and reachable set neglecting external environment constraints are obtained.

However, in order to guarantee controlled plant obstacle avoidance, the state trajectory should be kept outside all of spheres that represent obstacles. For achieving it, the invariant set S should not intersect with all of obstacle spheres. Suppose $S_0 \subset R^{n+p}$ is a sphere that is defined by $S_0=\{x(k) \mid x^T(k)Q_0 x(k) \le 1\}$, where $Q_0=\text{diag}[q_1, \ldots, q_{n+p}]$, $q_i=1/d_{\min}^2$, $d_{\min}$ is the distance between the center of the nearest obstacle and the current system equilibrium point.

According to the above analysis, the following theorem is proposed to calculate the maximum invariant set S and the reachable set S_r of system (4) under both external environment constraints and UH's limits.

Theorem 3: Given $\eta>0$, LQR parameter matrices Q_u, Q_x, symmetric positive-definite matrices W_0 and R, if there exist a symmetric positive-definite matrix $W \in R^{(n+p)\times(n+p)}$, matrices

$X \in R^{m_0 \times (n+p)}$, $Y \in R^{m_0 \times (n+p)}$, $Z \in R^{p \times m_0}$, diagonal positive-definite matrices $S \in R^{m_0 \times m_0}$, $W_r \in R^{p \times p}$, and a positive scale λ satisfying

$$\inf_{W,X,Y,Z,S,W_r} \lambda$$

$$\begin{bmatrix} 1 & X_i - Y_i \\ * & W \end{bmatrix} \geq 0, I = 1,\ldots,m_0$$

$$\begin{bmatrix} -W & Y^T & 0 & 0 & WA^T & -C^T & X^T & W \\ * & -2S & 0 & 0 & -SB^T - Z^T R^T & 0 & 0 & 0 \\ * & * & -\eta & 0 & D_1^T & 0 & 0 & 0 \\ * & * & * & -W_r & D_2^T & I & 0 & 0 \\ * & * & * & * & -W & 0 & 0 & 0 \\ * & * & * & * & * & -I & 0 & 0 \\ * & * & * & * & * & * & -Q_u^{-1} & 0 \\ * & * & * & * & * & * & * & -Q_x^{-1} \end{bmatrix} < 0$$

$$\begin{bmatrix} I_p & 0_{p \times n} \end{bmatrix}^T W_r \begin{bmatrix} I_p & 0_{p \times n} \end{bmatrix} \leq W$$

$$\begin{bmatrix} \lambda R & I \\ * & W_r \end{bmatrix} \leq 0$$

$$W \leq W_0$$

$$\lambda > 0$$

where $W_0 = Q_0^{-1}$, then obstacle avoidance and $-1 \leq v(k) \leq 1$ would be guaranteed, furthermore, the ellipsoid $\mathsf{S} = \{x(k) \mid x^T(k) P x(k) \leq 1\}$ is an invariant set with $P = W^{-1}$ and $\mathsf{S}_r = \{\text{ref} \mid \text{ref}^T Q_r \,\text{ref} \leq 1\}$ is a reachable set with $Q_r = W_r^{-1}$, which satisfies $\mathsf{S}_r \subset \mathsf{S}$.

Clearly, Theorem 2 is included in Theorem 3 because the invariant set is a bridge to connect environment and UH dynamic constraints. Thus, by the ISBP approach, the controller design and controller reference calculation are achieved simultaneously when the preknown work space, the desired path nodes, and the UH dynamic model are given.

The relationship between the related sets and the obstacles are shown in **Figure 5** assuming that all sets are projected to a 2D plane.

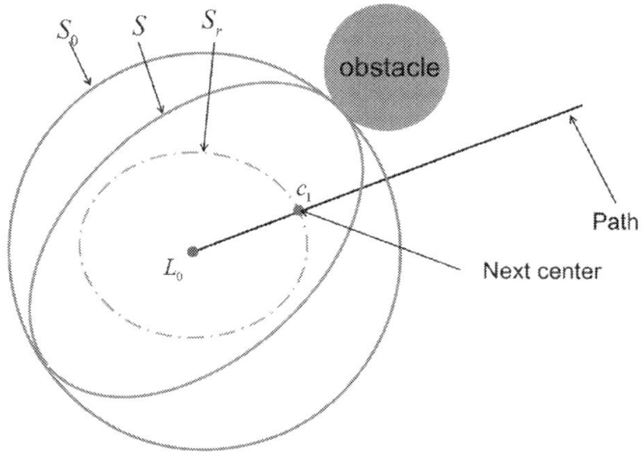

Figure 5. The relationship of related sets.

In other words, based on the calculated invariant set S and the reachable set S_r, a new center c_1 can be computed which is the intersection point of S_r and the feasible path. The next step is to calculate the second invariant set and the reachable set whose center is c_1 until the last reachable set covers the final path point.

6. Evaluation strategy

The open-loop model (2) of UH is unstable and the actuator outputs are limited. Thus, the only regional stability of UH can be guaranteed and the stability region is determined by the system structure and actuator constraints [16]. In other words, if the states of UH are inside the stability region, the UH would be of safety; otherwise, the UH may be in danger. Obviously, after actuator fault occurrence, actuator efficiency will be reduced. Hence, the safety region of the postfault system will be different with the fault-free case as shown in **Figure 6(a)**. Suppose that the safety region of the fault-free system with a fault-free controller is Ω_{ff} and the postfault system with a FTC controller is Ω_{pf} as shown in **Figure 6(b)**; furthermore, the initial state of the fault-free system is $x_0 \in \Omega_{\mathrm{ff}}$ and an actuator fault is detected at k_{f1}. Clearly, the state $x(k_{f1})$ is outside the safety region of the postfault system Ω_{pf} so that the postfault system may be unstable (as shown by state trajectory $x(k_{f1})x'(k_{f1})$) at last. That is to say, the actuator fault cannot be compensated. In contrast, suppose that the other fault is detected at k_{f2} and $x(k_{f2}) \in \Omega_{\mathrm{pf}}$ is valid, which represents that the fault can be compensated. However, the states in steady case are also determined by references such as x_{f1} which may be outside the safety region Ω_{pf}. Obviously, this reference is unreachable and tracking such reference may lead UH unsafe (as shown by the state trajectory $x(k_{f2})x'_{f1}$). Hence, reference reachability should be analyzed before system motion and a new reachable reference is necessary. Compared with x_{f1}, x_{f2} may be more reasonable which is inside Ω_{pf}.

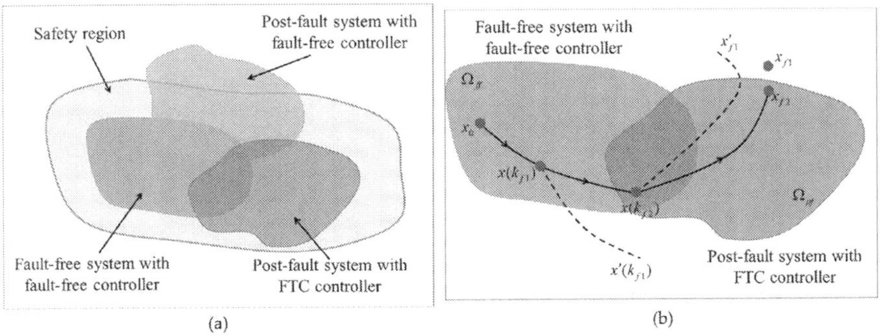

Figure 6. A sample of safety region of fault-free and postfault system in 2D state space.

According to Theorem 2, the safety region S is an invariant set. Clearly, a postfault system will be safe if initial states are inside the safety region S_f of the postfault system. In this way, the initial states of the postfault system can be evaluated. Second, the steady states will be analyzed. In steady-state case, actuators are not expected to be saturated so that the remaining efficiency of actuators can be used for disturbance defence. Hence, the original reference should be inside the reachable set of the postfault system such as ref $\in S_{rf}$, where S_{rf} is the reachable set of the postfault system; otherwise, the original reference ref is not reachable and tracking the original one may lead UH unsafe. The reason is that the actuator efficiency is reduced in the postfault system and tracking unreachable reference will lead fault-free actuator saturated which implies that UH cannot respond to control signal correctly. Under this condition, a new optimal reference is required which can be calculated by the trajectory replanning approach. In other words, if the original reference is not reachable after detecting the actuator fault, the ISBP approach should be called to calculate new trajectory and controller reference based on the postfault dynamic model of UH.

7. Numerical simulation

The parameters of nonlinear UH dynamic model used in numerical simulations is obtained from reference [6]. This model is used as the simulation model in this section. The step size of this simulation is 0.02s, which is the same as the control period of the real UH flight platform.

7.1. FDD simulation

Based on the simplified nonlinear model, the EKF is designed and LNN is used to train weight matrices and offsets of faulty actuators. The parameters of the EKF filter and the training results of the left swashplate actuator by LNN are as follows: $Q = diag(0, 0, 0, 0, 0, 0, 0, 0, 0, 0, 0)$, $R = 2.5\times 10^{-5}\cdot diag(40, 40, 40, 1, 1, 1, 1, 1, 1, 1, 1)$,
$W = [-0.0095, -0.00015, 0.000073, 0.076, 0.01023, -0.01008]$, $b = 0.03099$

where Q and R are covariance matrices of the process noise and the measurement noise, respectively. Then, we introduce a multiplicative fault on the left swashplate actuator as follows:

$$\begin{cases} \lambda_1 = 1 & 0 \le t < 20 \\ \lambda_1 = 0.6 & t \ge 20 \end{cases}$$

The fault detection module based on the filter residuals can get the WSSR curve shown in **Figure 7**. When no fault occurs, the WSSR curve cannot exceed the threshold.

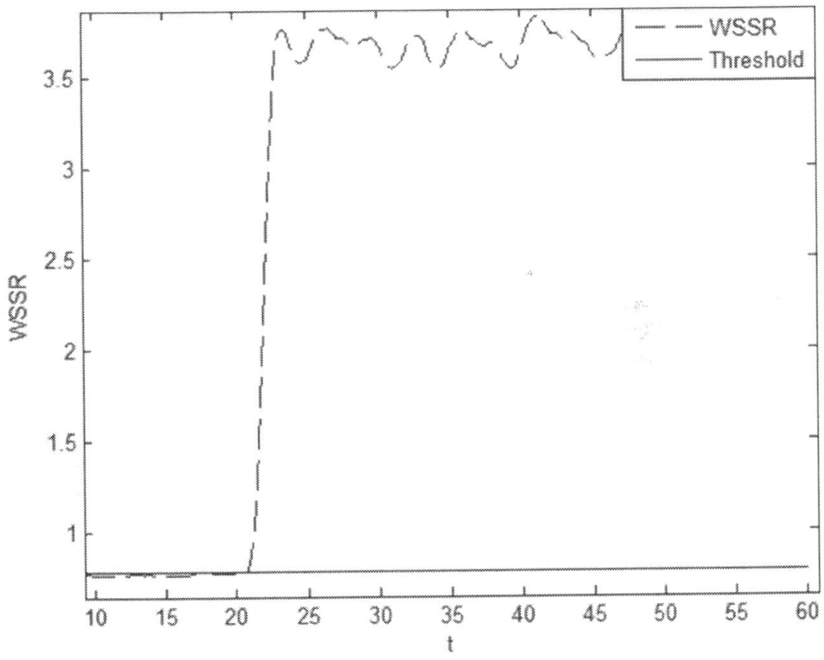

Figure 7. The curve of filter WSSR under the left servo fault.

After fault detection, isolation residuals are calculated to isolate this fault. The curves of four isolation residuals are shown in **Figure 8**. According to this figure, the isolation residual of the left swashplate actuator is less than other residuals. That is to say, the left actuator of swashplate is faulty. The fault isolation module cannot provide the right isolation results as soon as the completion of fault detection because of the vibration caused by the measurement noise. There may be misdiagnosis of the fault location within a very short time.

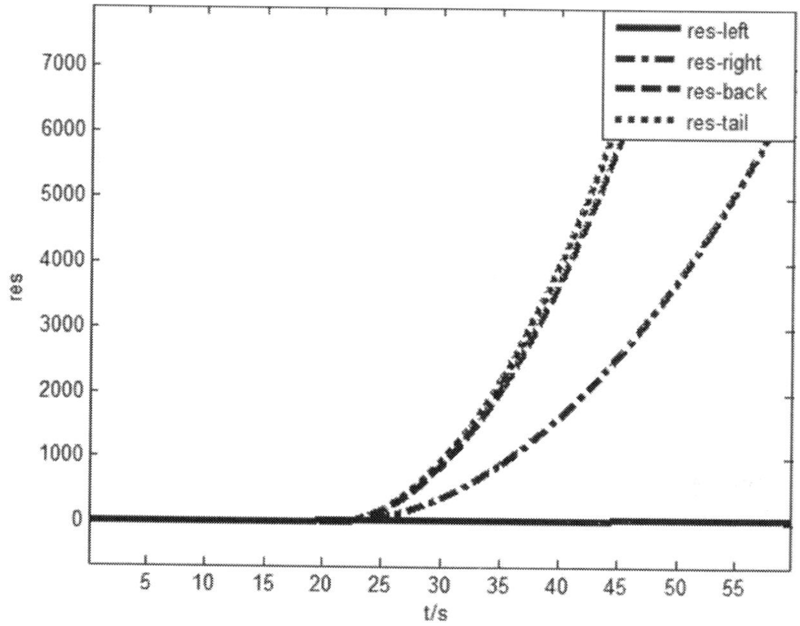

Figure 8. Curves of isolation residuals of every actuator under the left servo fault.

The curve of the proportional effectiveness of the left actuator under the fault is shown in **Figure 9**. After fault isolation, the isolation module passes the serial number of the faulty actuator to fault identification module to identify the AHC of this faulty actuator, and the simulation results indicate the size of this multiplicative fault. Because the weight matrices and offsets are trained by the LNN under the faults whose AHCs' parameters are all 0.5 for the online simulations, there may be error between the identification results and the actual size of this fault. By the calculation of average, the fault identification result is equal to the actual fault approximately.

The simulation results show that fault detection and isolation modules could detect and isolate a fault quickly, and the identification result is accurate enough; moreover, this method is with good real-time performance.

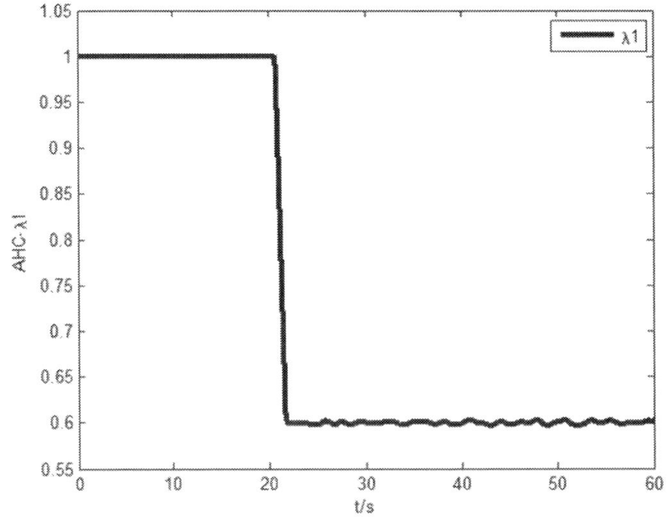

Figure 9. The curve of AHC proportional effectiveness under the left servo fault.

7.2. Controller and trajectory (re-)planning simulation

A linear UH model (2) with added three-axis positions is used in this part and related parameters can be found in [13]. The added positions are considered integration of velocities in the body coordinate. Thus, the system states, control inputs, and outputs are

$$x_h = [p_x \quad p_y \quad p_z \quad u \quad v \quad w \quad \phi \quad \theta \quad \psi \quad p \quad q \quad r \quad a_{1s} \quad b_{1s}]$$
$$u = [\theta_l \quad \theta_r \quad \theta_b \quad \theta_t \quad \omega_r],\ y_h = [p_x \quad p_y \quad p_z \quad \psi]$$

Assume that the left swashplate actuator has fault at 6.4s with λ_1=0.6, and the LQR parameter matrices are

$$Q_x = \text{diag}(0\ 0\ 0\ 0.01\ 0.01\ 0.01\ 3\ 1\ 0.02\ 0\ 0\ 0\ 0.1\ 0.1\ 0.001\ 0.001\ 0.003\ 0.01)$$
$$Q_u = \text{diag}(1110.50.1)$$

The preknown work space is shown in **Figure 10** where the start point is (0,0,0) and the target point is (20,20,20). The external environment constraint considered here is a sphere obstacle whose position is (10,10,10) and radius is 1. The feasible path is given by path nodes, such as (0,0,0), (8,8,13), and (20,20,20), which can guide UH from the start point to the target point with obstacle avoidance.

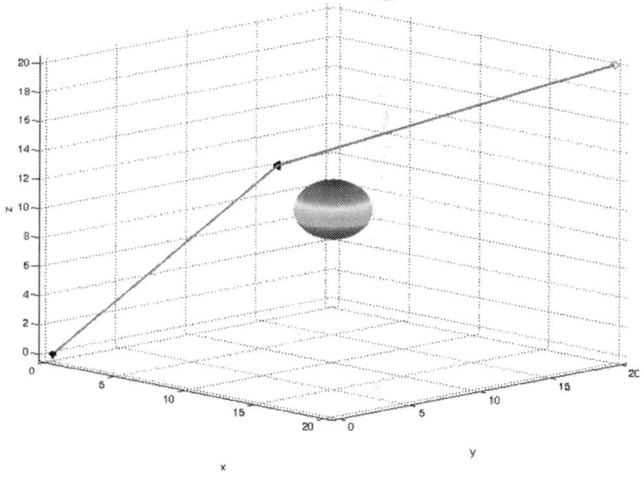

Figure 10. Preknown work space and feasible path.

According to Theorem 3, a series of controllers K, invariant sets S, and reachable sets S_r satisfying obstacle avoidance and UH dynamic limits can be computed as shown in **Figure 11** where the green ellipsoids are the reachable sets in 3D and the centers of ellipsoids, blue stars, are controller references. According to these references, the UH can run from the start point to the target point and the actual trajectory is shown by the blue curve. Clearly, the trajectory is kept inside the green ellipsoids so that it is also inside the invariant sets.

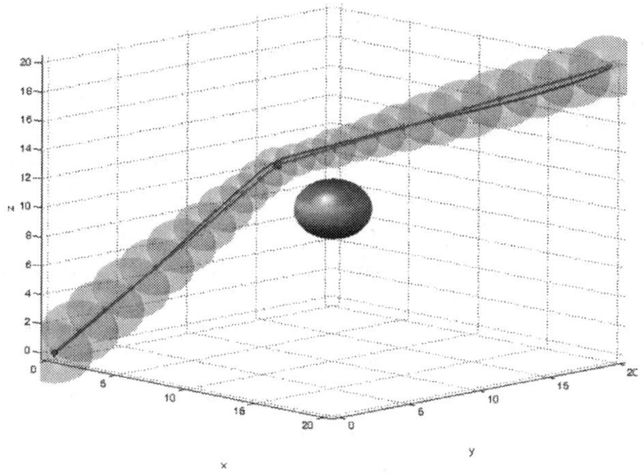

Figure 11. Reachable sets and UH actual trajectory in fault-free condition.

Compared with fault-free case as shown in **Figure 11**, **Figure 12** shows the results of the postfault case without the SHC framework. The actuator fault is detected at 6.4s and the related state is marked by red points in the figure. According to the postfault dynamic model, a new reachable set, red ellipsoid, is calculated. It is easy to see that the original reference, black point, is outside the reachable set of the postfault system which implies that the original reference is unreachable. Under this condition, if it does nothing, the UH system may be in danger as shown by the blue curve. The related manipulated variables are shown in **Figure 13** where the blue curves are the actuator outputs and the red dashed lines are the actuator constraints. Clearly, the actuator outputs are saturated at last which leads the UH system out of order.

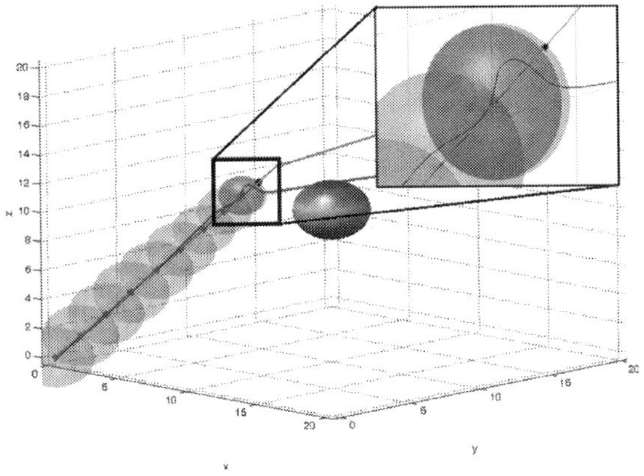

Figure 12. Reachable sets and UH actual trajectory in postfault condition without SHC.

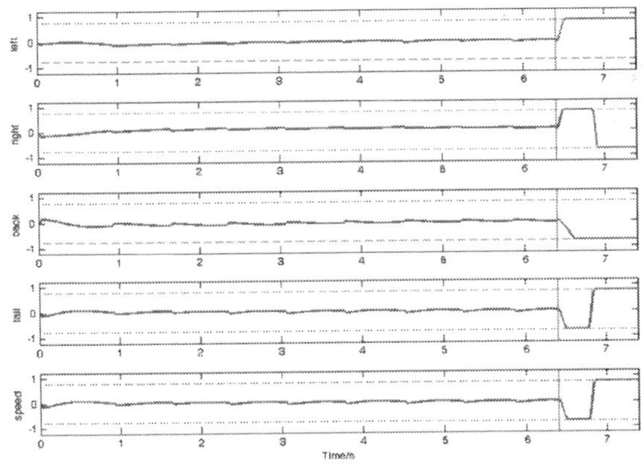

Figure 13. Actuator outputs in postfault condition without SHC.

Figure 14 shows the results of the postfault system with a SHC framework. After fault detection, Theorem 3 is recalled to calculate the new controllers, invariant sets, and reachable sets to evaluate the performance and guarantee the postfault UH system to be stable. At last, the UH can reach the target point with obstacle avoidance as shown by the actual trajectory.

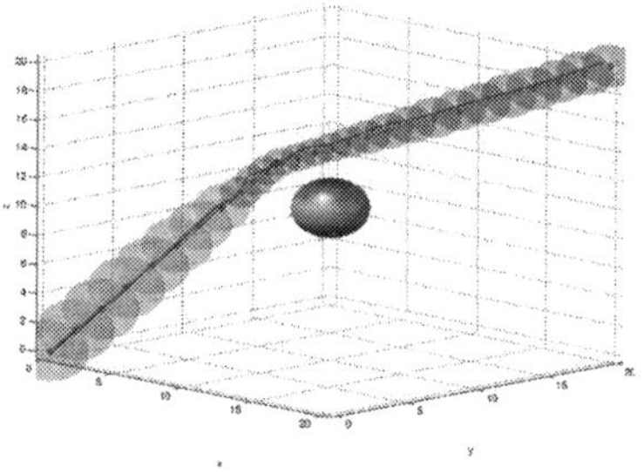

Figure 14. Reachable sets and UH actual trajectory in postfault condition with SHC.

8. Conclusions

In this chapter, a self-healing control framework is proposed for UH systems. The SHC framework aims at providing a solution to guarantee UHs safety and maximum ability to achieve the desired missions under both fault-free and postfault conditions. The EKF- and LNN-based FDD approach is used to detect and diagnosis actuator faults modeled by AHCs. Then, the AHC-based reconfigurable controller design method is proposed to calculate the fault-tolerant controller and the related safety region against both actuator faults and constraints by solving a set of LMIs. Third, the ISBP approach is presented for planning a feasible trajectory and computing the related controller reference under both external environment constraints and UH dynamic limits at the same time. After fault occurrence, based on the calculated safety region and controller reference, the performance of the postfault UH system can be evaluated, which can provide information whether the fault can be compensated and the original reference can be reached. If the original reference is not reachable, the ISBP approach will be recalled to calculate the new trajectory and reference again according to the postfault dynamic model. Finally, numerical simulations illustrate the effectiveness of the proposed SHC framework.

Acknowledgements

This work was supported by Key Program of National Natural Science Foundation of China under Grant 61433016 and National Natural Science Foundation of China under Grant 61503369.

Author details

Xin Qi[1,2], Zhong Liu[1,2], Yuqing He[1*], Liying Yang[1], Yuqing He[1] and Jianda Han[1]

*Address all correspondence to: heyuqing@sia.cn

1 Shenyang Institute of Automation Chinese Academy of Sciences, Shenyang, P.R.China

2 University of Chinese Academy of Sciences, Beijing, P.R.China

References

[1] Qi X., Qi J., Theilliol D., Zhang Y., Han J., Song D., Hua C. A review on fault diagnosis and fault tolerant control methods for single-rotor aerial vehicles. Journal of Intelligent & Robotic Systems. 2014;73(1):535–555. DOI: 10.1007/s10846-013-9954-z

[2] Qi J., Song D., Wu C., Han J., Wang T. KF-based adaptive UKF algorithm and its application for rotorcraft UAV actuator failure estimation. International Journal of Advanced Robotic Systems. 2012;9:1–9. DOI: 10.5772/51893

[3] Cai G., Chen B.M., Lee T.H. Unmanned Rotorcraft System. London: Springer; 2011. 270 pp. DOI: 10.1007/978-0-85729-635-1

[4] Shim H. Hierarchical flight control system synthesis for rotorcraft-based unmanned aerial vehicles [thesis]. Berkeley: University of California; 2000.

[5] Enns R., Si J. Helicopter flight-control reconfiguration for main rotor actuator failures. Journal of Guidance, Control, and Dynamics. 2003;26(4):572–584.

[6] He Y., Han J. Acceleration-feedback- enhanced robust control of an unmanned helicopter. Journal of Guidance, Control, and Dynamics. 2010;33(4):1236–1250.

[7] Ruschmann M., Wu N., Shin J.Y. Actuator fault diagnosis using two-stage extended Kalman filters. In: AIAA Guidance, Navigation, and Control Conference, 2–5 August, Toronto, Canada; 2010. pp. 1–21. DOI: 10.2514/6.2010-7703

[8] Ji H. Algebra foundation of the theory of control [dissertation]. Hefei: University of Science and Technology of China Press; 2008. 5 pp.

[9] Qiao C., Sun S. Fault detection for systems with missing measurements. In: The 25th Chinese Control and Decision Conference, 25–27 May, Guizhou, China; 2013. pp. 1050–1053.

[10] Blanchini F. Set invariance in control. Automatica. 1999;35(11):1747–1767. DOI: 10.1016/S0005-1098(99)00113-2

[11] Khalil H.K., Grizzle J.W. Nonlinear Systems. 3rd ed. New Jersey: Prentice- Hall; 2001. 750 pp.

[12] da Silva Jr. J.M.G., Tarbouriech S. Antiwindup design with guaranteed regions of stability: an LMI-based approach. IEEE Transactions on Automatic Control. 2005;50(1): 106–111. DOI: 10.1109/TAC.2004.841128

[13] Qi X., Qi J., Theilliol D., Song D., Zhang Y., Han J. Self-healing control design under actuator fault occurrence on single-rotor unmanned helicopters. Journal of Intelligent & Robotic Systems. 2016;online first,1–15. DOI: 10.1007/s10846-016-0341-4

[14] Qi X., Theilliol D., Song D., Han J. Invariant-set-based planning approach for obstacle avoidance under vehicle dynamic constraints. In: 2015 International Conference on Robotics and Biomimetics, IEEE, Zhuhai, China, 6–9 December,; 2015.

[15] Boyd S., Ghaoui E.L., Eric F., Venkataramanan B. Linear Matrix Inequalities in Systems and Control Theory. Philadelphia: SIAM; 1994. 205 pp.

[16] Hu T., Lin Z., Qiu L. An explicit description of null controllable regions of linear systems with saturating actuators. Systems & Control Letters. 2002;47(1):65–78. DOI: 10.1016/S0167-6911(02)00176-7

A Hierarchical Tracking Controller for Quadrotor Without Linear Velocity Measurements

Yassine Jmili, Nuradeen Fethalla,
Jawhar Ghomam and Maarouf Saad

Additional information is available at the end of the chapter

http://dx.doi.org/10.5772/62442

Abstract

This chapter deals with the position control of quadrotor unmanned aerial vehicle (UAV) when quadrotor's linear velocity is unavailable. We propose a hierarchical tracking controller for quadrotor UAV. The proposed controller does not require measurements of linear velocity of quadrotor. A nonlinear filter that avoids the need for measurements of linear velocity has proposed such that a global stability result is obtained for the position tracking error. However, backstepping based on barrier Lyapunov function has been used for the attitude controller. The control design is achieved by means of the hierarchical control, that is, design the position controller and attitude controller separately. This allows us to choose different nonlinear techniques for each controller.

Keywords: Quadrotor UAV, Tracking Control, Nonlinear Filter, Backstepping, Unknown linear velocity measurement

1. Introduction

In recent years, multirotor unmanned aerial vehicles (UAVs) have attracted many researchers and became a growing interest in the academic and industrial research. In this chapter, our study focuses on quadrotor UAVs due to its ability to fly in restricted areas. The quadrotor has four rotors controlled by four rotors. Quadrotors have specific characteristics that allow the execution of applications that would be difficult or complex compared with other flying machines. A key issue in making multirotor UAV vehicles feasible and efficient is the control design, consisting of a navigation system, position and attitude control. This chapter concerns the nonlinearity character of the UAV vehicles and focuses on the position and attitude controllers. One of scenarios that could be achieved by the quadrotor is autonomous navigation. Implementing such a mission requires the measurement of the components of vehicle state (position, linear velocity, orientation angles and angular velocity) from GPS sensors. Most of the control schemes that have been introduced in the literature rely on the available full state of feedback using definition of an observer or rely on the estimation of velocity of the vehicle by means of derivation of the successive measurements from the used sensors such GPS, etc. [1–3]. However, an additional disturbance may occur in the control loop as a result of the use of the observer. It is also important to validate first convergency of the observer. In addition, the stability of closed loop of complete system controlled by observer has to be assured through checking the compatibility of observer frequency and controller frequency [4]. In this estimation method, the errors produced by GPS may approach several meters, and in reality, velocity measurement error is growing fast by the induced measurement noise along with numerical integration. Few researchers have solved the problem of linear velocity estimation without the use of GPS, and some of them considered combination and artificial vision and the inertial sensors [5, 6].

In [7], the authors proposed a global exponential observer with a complete order to ensure a continued trajectory of a helicopter (VTOL). To reduce the size of the estimated state vector, a reduced observer introduced in [8], the linear velocity of the quadrotor has been estimated using a complete or partial measure of the acceleration such that the asymptotic stability of the error is obtained. Some other forms of observers have been introduced such as adaptive observer [9], and this approach provided an estimation of the velocity of quadrotor UAV from acceleration measurements calculated by the inertial system, but this method has disadvantages of slow convergence of the estimated parameters. While the sliding mode observer used in [10] has shown not only the ability to estimate the velocity of quadrotor but also showed its robustness against the external disturbances. The other technique used in the literature to estimate the linear velocity is the filters, and one of these filters is extended Kalman filter that widely applied to nonlinear dynamical systems. In this case, the principle is to use the standard equations of Kalman filter for nonlinear model linearized by Taylor formula of first order and the states of systems can be estimated from the noise measurements. These kinds of filters have been used to estimate the states of quadrotor [11].

In this paper, we propose an approach to solve the problem of position tracking of quadrotor UAV when the linear velocity measurements are unavailable. The proposed controller provides a hierarchical design of the system using assign inner loop for position control and outer loop for attitude control. Each loop has its own control algorithm. The resulting outer control loop, which is based on backstepping, has a simple structure. The inner control loop is based on nonlinear filter-based control design with the aim of estimating the linear velocity of quadrotor.

This paper is organized as follows. In Section 2, the dynamic model of quadrotor UAV is introduced. Section 3 presents the design of position controller based on the nonlinear dynamic model using nonlinear filter-based approach. In Section 4, the attitude controller of rotational system is presented. In Section 5, the overall closed loop system stability is analysed, and simulation results are shown in Section 6.

http://dx.doi.org/10.5772/62442

2. Quadrotor dynamic model

The quadrotor UAV is modelled as a rigid body of mass $m \in R$ and of matrix of inertia $I \in R^{3x3}(I = dia(I_1, I_2, I_3))$. The equations of motion can be described by two reference frames: first one is the reference frame (I) combined with the unit vector basis (e_1, e_2, e_3) and second one is the body frame (B) fixed on the quadrotor UAV and combined with vector basis (e_1^b, e_2^b, e_3^b) (see **Figure 1**). The kinematic and dynamic equations of motion for the quadrotor expressed in the body-fixed reference frame can be described as [12]. The dynamic model is derived using Newton-Euler formalism in the body fixed frame B.

$$\begin{aligned} \dot{\eta} &= RV \\ \dot{V} &= -s(\Omega)V + \frac{1}{m} f + \frac{1}{m} R^T b_v \\ \dot{\Theta} &= k(\Theta)\Omega \\ I_f \dot{\Omega} &= -\Omega \times I_f \Omega - G_a + \tau_a \end{aligned} \tag{1}$$

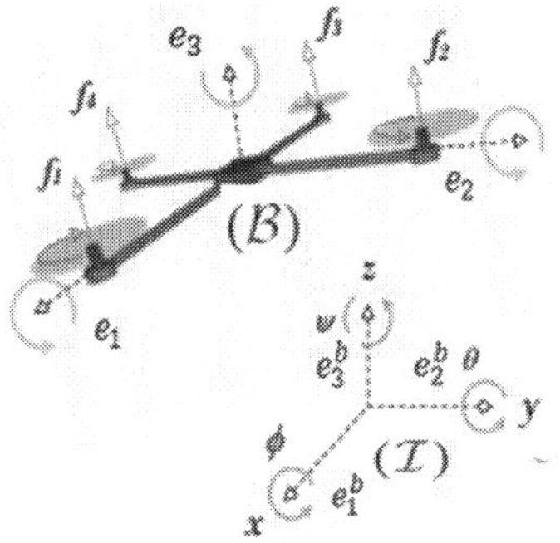

Figure 1. Quadrotor airframe and reference frames.

where $\eta = [x, y, z]^T$ represents the position of the centre of mass of the quadrotor, these positions are expressed relatively with respect to an inertial frame and $\Theta = [\psi, \theta, \varphi]$ represents the three Euler angles (yaw, pitch, roll) that define the orientation of the quadrotor in space. $V = [v_x, v_y, v_z]^T \in R^3$ and $\Omega \in R^3$ denote the linear and angular velocities expressed in the body-fixed frame B. b_v represents the external force disturbance expressed in the inertial frame I. m is the mass of the quadrotor, $I_f \in R^{3\times3}$ denotes a positive definite inertia matrix relative to the inertial frame I. The vector $f, \tau_a \in R^3$ denotes the external force and torque expressed in B. The external force is aligned with the $z-axis$ and can be written as $f = -G + Te_3$, where $G \in R^3$ is the gravity vector defined as $G = mgR^T e_3$ where $g \in R$ denotes the gravitational acceleration and

$e_3=[0,0,1]^T$ is the unit vector in the inertial frame. T denotes the total thrust produced by the four rotors given as $T=k\sum_1^4\omega_i^2$ where ω_i being the speed of the rotor i, k is the thrust factor and it is a positive proportional constant parameter depending on the density of air, the radius, the shape, the pitch angle of the blade and other factors. $G_a \in R^3$ contains gyroscopic couples, due to the rotational motion of the quadrotor and the four rotors which is given by $G_a=\sum_1^4 I_r(\Omega e_3)(-1)^{i+1}\omega_i$ where I_r denotes the moment of inertia. The generalized moments on the Θ variable are as follows.

$$\tau_a = \begin{pmatrix} \tau_\psi \\ \tau_\theta \\ \tau_\phi \end{pmatrix} = \begin{pmatrix} k(\omega_1^2 - \omega_2^2 + \omega_3^2 - \omega_4^2) \\ kl(\omega_2^2 - \omega_4^2) \\ kl(\omega_3^2 - \omega_1^2) \end{pmatrix} \tag{2}$$

where l denotes the distance from the rotor to the centre of mass of the quadrotor. Using Euler angles parameterization and "XYZ" convention, the airframe orientation in space is given by a rotation matrix R from B to I, where $R \in SO_3$ is expressed as follows:

$$R = R_\psi R_\theta R_\phi$$
$$\begin{pmatrix} C_\theta C_\psi & S_\phi S_\theta C_\psi - C_\phi S_\psi & C_\phi S_\theta C_\psi + S_\phi S_\psi \\ C_\theta S_\psi & S_\phi S_\theta S_\psi + C_\phi C_\psi & C_\phi S_\theta S_\psi - S_\phi S_\psi \\ -S_\theta & S_\phi C_\theta & C_\phi C_\theta \end{pmatrix} \tag{3}$$

3. Quadrotor forces and moments

The four rotors of quadrotor, rotating at angular velocities $omega_i$, produce the four forces $f_i=k\omega_i^2$, pointed upward, where i = 1,2,3,4, and k_i are positive constants. These four forces constitute the total thrust T along z axis. Therefore, the distributed forces and moments from four actuator motors for the quadrotor are given by:

$$\begin{bmatrix} T \\ \tau_\psi \\ \tau_\theta \\ \tau_\phi \end{bmatrix} = \begin{bmatrix} 1 & 1 & 1 & 1 \\ -l & 0 & l & 0 \\ 0 & l & 0 & -l \\ c & -c & c & -c \end{bmatrix} \begin{bmatrix} f_1 \\ f_2 \\ f_3 \\ f_4 \end{bmatrix} \tag{4}$$

where c is the drag coefficient. Since the quadrotor UAV is under-actuated system, the design of control inputs for this kind of systems is a challenging topic.

4. Control problem formulation

In this work, we aim to design control laws for the total thrust T and torques τ_a allowing the quadrotor UAV to track a desired trajectory $\eta_d = [x_d, y_d, z_d]$ and desired heading ψ_d. The set of trajectory and their derivatives are smooth enough so that they are uniformly continuous and bounded. The linear velocity is assumed to be not available for feedback and only the position and acceleration are therefore accessible by the translational subsystem. However, we want to design a global feedback control law in the absence of measurements of linear velocity that guarantee the position, heading and velocity tracking errors that are bounded and converge asymptotically to zero.

$$e_\psi = \psi - \psi_d, e_\eta = \eta - \eta_d, \dot{e}_\eta = v - v_d \tag{5}$$

The main goal of this paper is to solve the position control problem; meanwhile, guaranteeing all closed-loop signals are bounded and the corresponding constraints of the attitude system are violated. In the following, we will introduce the methodology of control design to achieve our objectives.

5. Control design procedure

The equations of motion (2) of quadrotor defined in the previous section can be rewritten as

$$\Sigma 1 \begin{cases} \dot{\eta} = v \\ \dot{v} = \dfrac{1}{m} Re_3 T - g e_z \end{cases} \tag{6}$$

$$\Sigma 2 \begin{cases} \dot{R}e_3 = RS(\Omega)e_3 \\ \dot{\Theta} = K(\Theta)\Omega \\ \dot{\Omega} = -\Omega \times (I\Omega) - G_a + \tau_a \end{cases} \tag{7}$$

The dynamic model above is separated into two dynamic systems; translational dynamic system Σ_1 and rotational dynamic system Σ_2. These two systems are chosen completely coupled by the rotation vector Re_3 and total thrust vector T as shown in **Figure 2**. In our methodology, we adopt a hierarchical controller design for the entire system dividing it into translational and rotational dynamics subsytems as it is shown in **Figure 2**. In this design, we assign the outer loop for the position controller and the inner loop for the attitude controller. The control design procedure can be summarized in the following two points:

1. In the design of the position controller, we propose a nonlinear filter to estimate the linear velocity tracking error rather than having its feedback measurement and design the

intermediate position control from which we can calculate the desired value of the total thrust and also extract the desired orientation that will be considered as a reference for the rotational dynamics loop.

2. For the attitude controller, backstepping technique and control design with a barrier Lyapunov function have been applied due to the fact that the dynamic of Σ_2 is in the form of "strict feedback" system, which is appropriate for such a controller. The virtual control laws for angular velocities resulting from applying backstepping technique are obtained $\alpha_\Omega = [\alpha_r, \alpha_p, \alpha_q]$ to ensure the asymptotic convergence of followed rotation angles errors to zero, from the torques. Then, the four rotational speed signals can be calculated by the transformation matrix R.

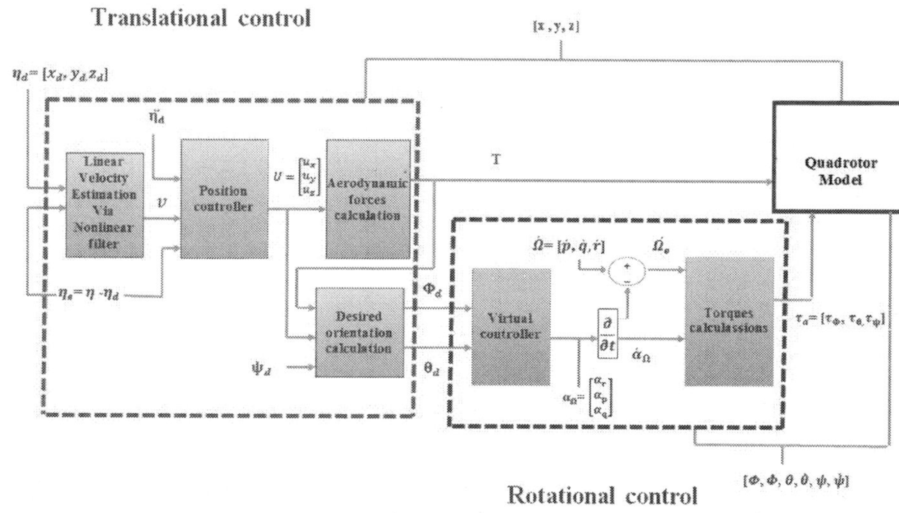

Figure 2. Control architecture.

We will introduce barrier Lyapunov function (BLF) to design the attitude tracking controller. This function is designed to avoid the constraints (singularity points) that may occur in the rotation angles, roll, pitch and yaw and to produce a feasible initial condition for torques control.

Definition. A barrier Lyapunov function is a scalar function $V(x)$, defined with respect to the system $\dot{x} = f(x)$ on an open region D containing the origin, that is continuous, positive definite, has continuous first-order partial derivative at every point of D, has the property $V(x) \to \infty$ as x approaches the boundary of D and satisfies $V(x(t)) \leq b\ \forall\ t \geq 0$ along the solution of $\dot{x} = f(x)$ for $x(0) \in$ D and some positive constant b [13].

Lemma 1. For any positive constant k_{b_1}, let $Z_1 = z_1 \in R : z_1 < |k_1| \subset R$ and $N := R^l \times Z_1 \subset R^{l+1}$ be open sets [13].

Consider the system

$$\dot{\zeta} = h(t, \zeta) \tag{8}$$

where $\zeta := [\omega, z_1]^T \in N$, and $h : R_+ \times N \to R^{l+1}$ is piecewise continuous in t and locally Lipschitz in z, uniformly in t, on $R_+ \times N$. Suppose that there exist functions $U : N^l \to R_+$ and $V_1 : Z_1 \to R_+$, continuously differentiable and positive definite in their respective domains, such that

$$V_1(z_1) \to \infty as z_1 \to z_1 \to \pm k_{b_1} \tag{9}$$

$$\sigma_1(\|\omega\|) \le U(\omega) \le \sigma_2(\|\omega\|) \tag{10}$$

where σ_1 and σ_2 are class k_∞ functions. Let $V(\zeta) := V_1(z_1) + U(\omega)$ and $z_1(0)$ belong to the set $z_1 \in (-k_{b_1}, k_{b_1})$. If the inequality holds:

$$\dot{V} = \frac{\partial V}{\partial \zeta} h \le 0 \tag{11}$$

then $z_1(t)$ remains in the open set $z_1 \in (-k_{b_1}, k_{b_1})\ \forall\ t \in [0, \infty)$.

Proof. The conditions on h ensure the existence and uniqueness of a maximal solution $\zeta(t)$ on the time interval $[0, \tau_{max})$, according to [14]. This implies that $V(\zeta(t))$ exist for all $t \in [0, \tau_{max})$.

Since $V(\zeta)$ is positive definite and $\dot{V} \le 0$, we know that $V(\zeta(t)) \le V(\zeta(0))$ for all $t \in [0, \tau_{max})$. From $V(\zeta) := V_1(z_1) + U(\omega)$ and the fact that $V_1(z_1)$ and $U(\omega)$ are positive functions, it is obvious that $V_1(z_1(t))$ is also bounded for all $t \in [0, \tau_{max})$. As a result, it is known from Eq. (9) that $|z_1| \ne \pm k_{b_1}$. Based on the $z_1(0) < |k_{b_1}|$, We can figure out that $z_1(t)$ remains in the set $z_1 < |k_{b_1}|$ for all $t \in [0, \tau_{max})$.

Accordingly, there exist a compact set $K \subseteq N$ such that the maximal solution of Eq. (8) satisfies $\zeta(t) \in K$ for all $t \in [0, \tau_{max})$. Based on [14], we conclude that $\zeta(t)$ is defined for all $t \in [0, \infty)$ and hence $z_1(t) \in (-k_{b_1}, k_{b_1})\ \forall\ t \in [0, \infty)$.

5.1. Position control design

To design the position control for the dynamics of translational subsystem Eq. (6) that accomplish our control goal when no linear velocity measurements are available, we have considered a nonlinear first-order filter to estimate the tracking error of the linear velocity. The classification of the designed filter is output feedback controller [15]. It is designed with hyperbolic tangent function to ensure boundedness of thrust vector. The form of the filter can be written as

$$\begin{aligned} v &= q + B\eta_e \\ \dot{q} &= -A\tanh(v) \end{aligned} \tag{12}$$

where A and B are two diagonal matrices.

Next section presents an explanation of operation of selected filter and simple simulation example.

5.1.1. Filter principal design

The main idea of our designed filter is inspired from a simple first-order linear high-pass filter [16].

$$\frac{q(j\omega)}{x_1(j\omega)} = \frac{\beta}{j\omega - \beta} \tag{13}$$

To interpret the filter in frequency domain, we introduce the variable $s = j\omega$ and replace this variable in Eq. (13), which yields

$$sq = \beta(x_1 + q) \tag{14}$$

where

$$q = \frac{\beta x_1}{s - \beta}$$

extracting the term $(x_1 + q)$, we get

$$(x_1 + q) = x_1 + \frac{\beta x_1}{s - \beta} = \frac{s x_1}{s - \beta} \tag{15}$$

5.1.2. Intermediate position control design

The goal of this part is to design an intermediate control that ensures the convergence of position tracking and velocity errors to zero.

$$\lim_{t\to 0}\|\eta-\eta_d\|=0,\ \lim_{t\to 0}\|v-v_d\|=0 \tag{16}$$

The choice of vector input $U=[u_x,u_y,u_z]^T=\frac{1}{m}Re_3T$ leads easily to extract the desired orientation since it constitutes the link between the position dynamics and attitude dynamics. This can be formulated as

$$U=\frac{1}{m}Re_3T=-k_p\tanh(\eta_e)-k_d\tanh(v)+\ddot{\eta}_d+ge_z \tag{17}$$

where k_p and k_d are two column vectors that contain three elements which will be defined by the proof of stability of position controller. The boundedness of Eq. (17) must be taken into consideration.

$$\|U\|=\|\frac{T}{m}\|=\|-k_p\tanh(\eta_e)-k_d\tanh(v)+\ddot{\eta}_d+ge_z\| \tag{18}$$

considering

$$\|U\|=\|\frac{T}{m}\|\le k_p+k_g+k_d \tag{19}$$

such that: $k_g\le\|\ddot{\eta}_d\|+g$. So, the selected positive variables of k_p,k_d, k_g guaranteeing the vector thrust to be bounded.

To show the stability of closed loop position system with the controller Eq. (19), we will introduce it later in the Appendix A.

5.2. Attitude control design

The control of the attitude system is implemented to ensure the asymptotic convergence of desired orientation. we exploited the two interconnected blocks: block of virtual control $\alpha_\Omega=[\alpha_r,\alpha_p,\alpha_q]^T$ and block of torque control $\tau_a=[\tau_\varphi,\tau_\theta,\tau_\psi]$. Consider the subsystem (6, Σ_2)

$$\dot{\Theta}=k(\Theta)\Omega \tag{20}$$

which describes the rotation system, the angular velocity vector Ω and torque vector τ_a that can be restated as

$$\Sigma_2=\begin{cases}\dot{\Theta}=f_1(\Omega)\\ \dot{\Omega}=f_1(\Omega,\tau)\end{cases} \tag{21}$$

Thus, the virtual angular velocity control laws were introduced to produce a zero error of the derivative of angular position $\dot{\Theta}=[\dot{e}_{\varphi},\dot{e}_{\theta},\dot{e}_{\psi}]$ and to obtain the appropriate torque for the control system. From Eq. (20), we define the angular velocity tracking error vector $e_{\Omega}=[e_p,e_q,e_r]^T$. where

$$e_{\Omega} = \Omega - \alpha_{\Omega} \tag{22}$$

By substituting the Eq. (22) into Eq. (20) yields

$$\begin{bmatrix} \dot{\phi} \\ \dot{\theta} \\ \dot{\psi} \end{bmatrix} = \begin{bmatrix} 1 & \sin(\phi)\tan(\theta) & \cos(\phi)\tanh(\theta) \\ 0 & \cos(\phi) & -\sin(\phi) \\ 0 & \dfrac{\sin(\phi)}{\cos(\theta)} & \dfrac{\cos(\phi)}{\cos(\theta)} \end{bmatrix} \begin{bmatrix} e_p + \alpha_p \\ e_q + \alpha_q \\ e_r + \alpha_r \end{bmatrix} \tag{23}$$

similarly,

$$\begin{cases} \dot{\phi} = e_p + \alpha_p + \sin(\phi)\tan(\theta)q + \cos(\phi)\tan(\theta)r \\ \dot{\theta} = \cos(\phi)(e_q + \alpha_q) - \sin(\phi)r \\ \dot{\psi} = \dfrac{\cos(\phi)}{\cos(\theta)}(e_r + \alpha_r) + \dfrac{\sin(\phi)}{\cos(\theta)}q \end{cases} \tag{24}$$

the derivative of angular position error can be derived below

$$\begin{aligned} \dot{e}_{\phi} &= \dot{\phi} - \dot{\phi}_d \\ &= e_p + \alpha_p + \sin(\phi)\tan(\theta)q + \cos(\phi)\tan(\theta)r - \dot{\phi}_d \\ \dot{e}_{\theta} &= \dot{\theta} - \dot{\theta}_d \\ &= \cos(\phi)(e_q + \alpha_q) - \sin(\phi)r\dot{\theta}_d \\ \dot{e}_{\psi} &= \dot{\psi} - \dot{\psi}_d \\ &= \frac{\cos(\phi)}{\cos(\theta)}(e_r + \alpha_r) + \frac{\sin(\phi)}{\cos(\theta)}q - \dot{\psi}_d \end{aligned} \tag{25}$$

We choose an appropriate barrier Lyapunov function for roll and pith angles. And this is not needed for the yaw angle since it has singularity-free condition and we can use a single quadratic function to derive its control law.

The augmented Lyapunov function can be stated as:

$$V_1 = \frac{1}{2}e_\psi{}^2 + \frac{1}{2}\log\frac{k_\phi{}^2}{k_\phi{}^2 - e_\phi{}^2} + \frac{1}{2}\log\frac{k_\theta{}^2}{k_\theta^2 - e_\theta^2} \tag{26}$$

and its time derivative is

$$\dot{V}_1 = e_\psi \dot{e}_\psi + \frac{e_\phi \dot{e}_\phi}{k_\phi{}^2 - e_\phi{}^2} + \frac{e_\theta \dot{e}_\theta}{k_\theta^2 - e_\theta^2} \tag{27}$$

So, the virtual control law for the angular velocity is derived as follows:

$$\begin{aligned} \alpha_p &= -\sin(\phi)\tan(\theta)q - \cos(\phi)\tan(\theta)r + \dot{\phi}_d - k_\phi e_\phi \\ \alpha_q &= -k_\theta \cos(\phi)e_\theta + \tan(\theta)r + \frac{\dot{\theta}_d}{\cos(\phi)} \\ \alpha_r &= \frac{\cos(\theta)}{\cos(\phi)}(\dot{\psi}_d - \frac{\sin(\phi)}{\cos(\theta)}q - k_\psi e_\psi) \end{aligned} \tag{28}$$

Substituting Eq. (28) into Eq. (25) and then substituting the resulting equations into (27) yields

$$\begin{aligned} \dot{V}_1 &= -k_\psi e_\psi^2 - \frac{k_\phi e_\phi^2}{k_\phi^2 - e_\phi^2} \\ &- \frac{k_\theta \cos^2(\phi) e_\phi^2}{k_\theta^2 - e_\theta^2} + \frac{\cos(\phi)}{\cos(\theta)} e_r e_\psi \\ &+ \frac{e_\theta \cos(\phi) e_q}{k_\theta^2 - e_\theta^2} + \frac{e_\phi e_p}{k_\phi^2 - e_\phi^2} \end{aligned} \tag{29}$$

It is obvious that Eq. (29) has negative terms and these terms can stabilize the system only if the conditions $|e_\varphi| < k_\varphi, |e_\theta| < k_\theta \forall k_\varphi > 0 and K_\theta > 0$ are met. These conditions will be met using Lemma (1). The other positive terms have to be cancelled and this will be done in the next. Now, backstepping control algorithm was applied to the partial of rotation system to control the input torque τ_a. The stability of rotation closed loop system is proved. We define the angular velocity error and its derivative:

$$\Omega_e = \Omega - \alpha_\Omega \tag{30}$$

$$\dot{\Omega}_e = \dot{\Omega} - \dot{\alpha}_\Omega \tag{31}$$

Consider the following positive definite Lyapunov function candidate:

$$V_{BF} = V_1 + \frac{1}{2}\Omega_e^T I \Omega_e \tag{32}$$

and its time derivative is

$$\dot{V}_{BF} = \dot{V}_1 + \Omega_e^T I \dot{\Omega}_e \tag{33}$$

Substituting Eq. (30) into Eq. (32) yields

$$\dot{V}_{BF} = \dot{V}_1 + \Omega_e^T [I\dot{\Omega} - I\dot{\alpha}_\Omega] \tag{34}$$

Replacing the third equation of system (6, Σ_2) in Eq. (33) yields

$$\dot{V}_{BF} = \dot{V}_1 + \Omega_e^T [-\Omega \times I\Omega - G_a + \tau_a - I\dot{\alpha}_\Omega] \tag{35}$$

The time derivative of V_{bf} ,using Eqs. (29) and (34), is given by

$$\dot{V}_{BF} = \dot{V}_1 + \Omega_e^T [-(\Omega_e + \alpha_\Omega) \times I\Omega - G_a + \tau_a - I\dot{\alpha}_\Omega] \tag{36}$$

In view of last equation, we propose the following torque input for the rotational dynamics

$$\begin{aligned} \tau_a &= k_e \Omega_e - \frac{e_\phi}{k_\phi^2 - e_\phi^2}\begin{pmatrix} 0 \\ 1 \\ 0 \end{pmatrix} - \frac{\cos(\phi) e_\theta}{k_\theta^2 - e_\theta^2}\begin{pmatrix} 0 \\ 0 \\ 1 \end{pmatrix} \\ &- \frac{\cos(\phi)}{\cos(\theta)} e_r e_\psi + G_a + I\dot{\phi}_\Omega + \phi_\Omega I\Omega \end{aligned} \tag{37}$$

and hence

$$\begin{aligned} \dot{V}_1 &= -k_\psi e_\psi^2 - \frac{k_\phi e_\phi^2}{k_\phi^2 - e_\phi^2} \\ &- \frac{k_\theta \cos(\phi) e_\theta^2}{k_\theta^2 - e_\theta^2} \\ &- \Omega_e^T k_e \Omega_e - \Omega_e^T \Omega_e I\Omega \le 0 \end{aligned} \tag{38}$$

6. Numerical simulation results

To show the effectiveness of the proposed control algorithm, we have introduced the simulation results. In order to test the overall system using Matlab simulink, the quadrotor UAV modelled as a rigid body of mass $m=1.169Kg$ with inertial matrix $I=dia(col(0.024,0.024,0.032))Kg.m^2$, $k=1.41\times10^{-5}$, $c=1.84\times10^{-6}$, $I_r=7.81\times10^{-5}Kg.m^2$ and $l=0.2125m$. The numerical simulation results have been obtained using Runge-Kutta's method with variable time step 0.001s. The desired trajectory is given by $x_d(t)=0,\ \ y_d t=0,\ \ z_d(t)=-(1-e^{\frac{t}{10}})$, $\psi_d(t)=\frac{\pi}{4}\,for 20\leq t\leq 30$with initial conditions $x(0)=0,\ y()=0,\ z(0)=0,\ \psi(0)=0$. The controller gains are as follows: $A=[40,40,40]^T, B=[30,30,30]^T, k_p=[160,160,160]^T, k_d=[20,20,20]^T, k_\varphi=4, k_\theta=8, k_\psi 4$, $k_e=[7,7,7]^T$. The obtained results are shown in **Figures 3, 4, 5, 6, 7, 8** and **9**.

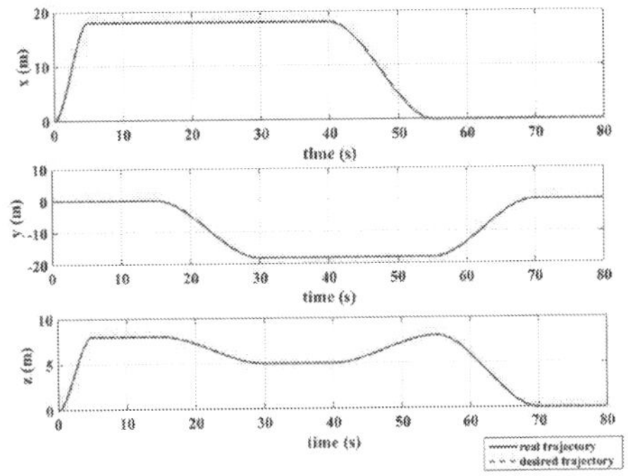

Figure 3. Position tracking.

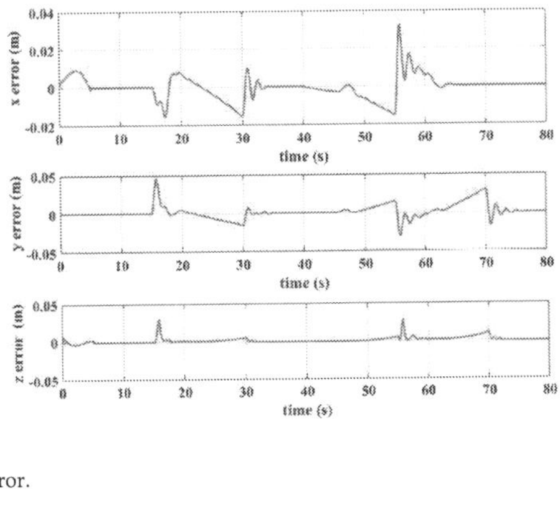

Figure 4. Position tracking error.

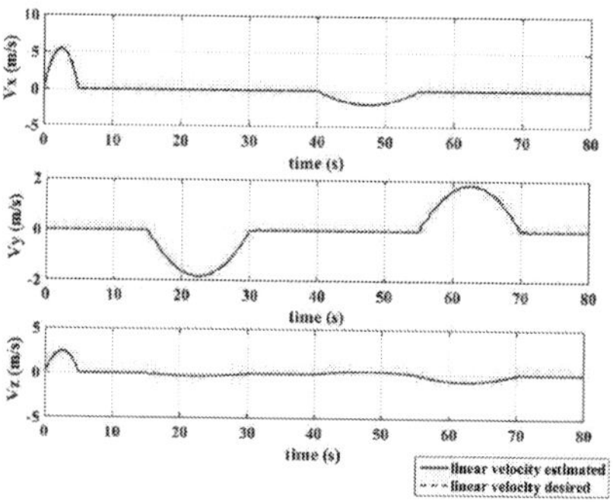

Figure 5. Estimated velocity tracking.

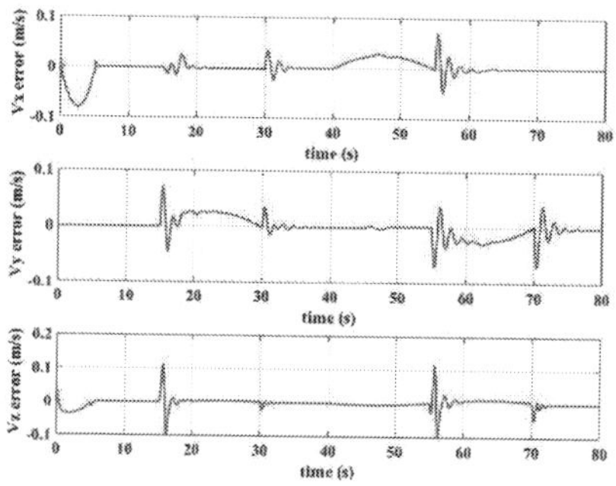

Figure 6. Estimated velocity tracking error.

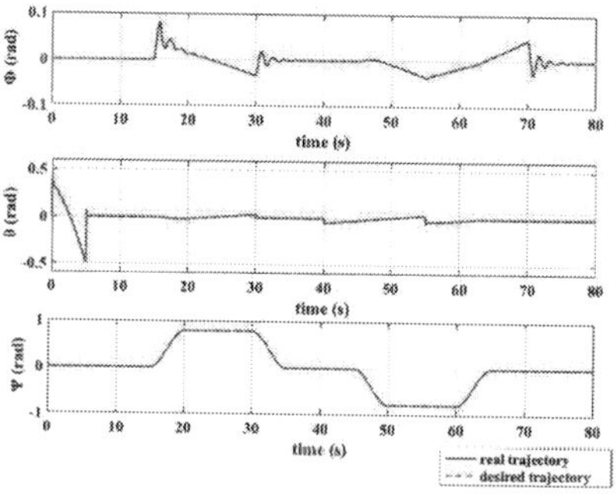

Figure 7. Attitude tracking.

http://dx.doi.org/10.5772/62442

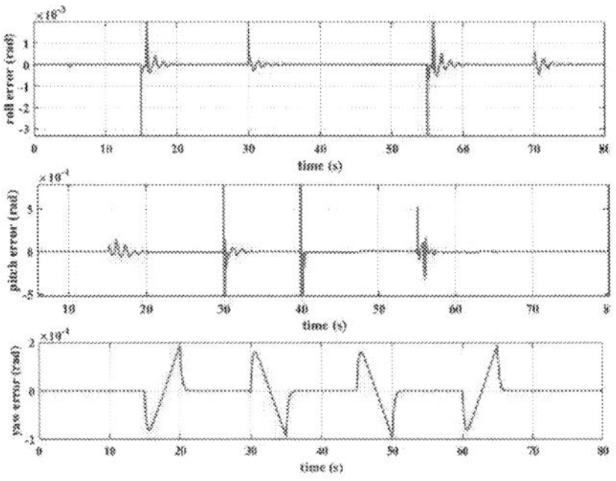

Figure 8. Attitude tracking error.

Figures 3, 4, 5 and **6** show the three components of the position and estimated velocity tracking errors. **Figures 7** and **8** illustrate the attitude tracking and its tracking errors. To show the quadrotor's position tracking, a 3D plot of the quadrotor's position with desired trajectory is given in **Figure 9**.

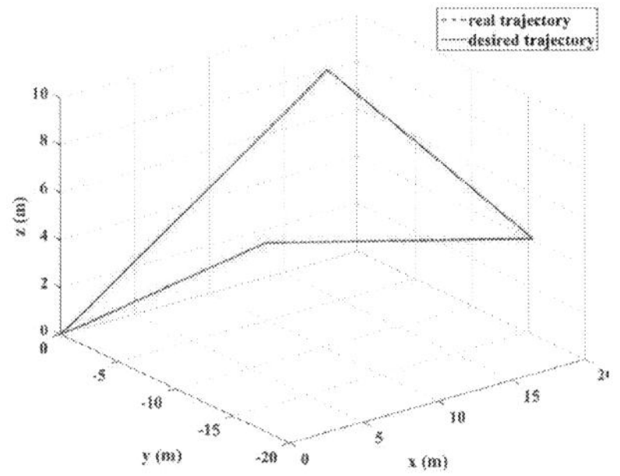

Figure 9. 3D trajectory of the quadrotor UAV (dashed blue) with the desired trajectory (red).

7. Conclusion

In this work, we addressed the trajectory tracking problem of quadrotor UAV without linear velocity measurements. The presented algorithm was based on a separate translational and rotational control design, and the overall closed loop system showed a global asymptotic stability. Exploited the nature structure of the model, the controller is designed in a hierarchical manner by considering the separation between the translational and orientation dynamics. In the first phase of our control design, the linear velocity measurement has been avoided with the introduced nonlinear filter without using any observer. During this phase, the position controller computes the total thrust, considered as input for translational dynamics, and the desired orientation. In the second phase of our control design, barrier Lyapunov function has been employed to design the control torque that ensuring the tracking of the desired attitude derived at first phase of the control design. The stability of the closed loop system has been investigated and the simulation results have been provided to show the good performance of the proposed algorithm. Our next work will be to accomplish real time experiments for the proposed algorithm.

Appendix

Stability analysis of position control without velocity measurements

Considering the dynamic equations of translational system (6, Σ_1), the dynamics of acceleration error can be rewritten as

$$\ddot{\eta}_e = \frac{1}{m} Re_3 T - ge_3 - \ddot{\eta}_d \tag{39}$$

The closed loop of position system is obtained by substituting the control intermediate vector U to the dynamics of acceleration.

$$S_{BF_1} \begin{cases} \ddot{\eta}_e = -k_p \tanh(\eta_e) - k_d \tanh(v) \\ \dot{v} = -A \tanh(v) + B\dot{\eta}_e \end{cases} \tag{40}$$

Consider the following candidate Lyapunov function

$$\begin{aligned} V_{BF_1} &= \frac{1}{2}\dot{\eta}_e^T \dot{\eta}_e + k_p \log(\cosh(\eta_e)) \\ &+ k_p B^{-1} \log(\cosh(V)) + \gamma\dot{\eta}_e (\tanh(\eta_e) - \tanh(v)) \end{aligned} \tag{41}$$

where γ is a small positive parameter to be determined later. Taking the time derivative of Eq. (41) yields

$$\begin{aligned} \dot{V}_{BF_1} &= \frac{1}{2}\dot{\eta}_e{}^T \ddot{\eta}_e + k_p \tanh(\eta_e^T)\dot{\eta}_e \\ &+ k_d B^{-1} \tanh(v^T)\dot{v} + \gamma\ddot{\eta}_e (\tanh(\eta_e) - \tanh(v)) \\ &+ \gamma\dot{\eta}_e^T (\cosh^{-2}(\eta_e)\dot{\eta}_e - \cosh^{-2}(v)\dot{v}) \end{aligned} \tag{42}$$

substituting Eqs. (40)–(42) yields

http://dx.doi.org/10.5772/62442

$$\begin{aligned}\dot{V}_{BF_1} &= -k_d B^{-1} A \tanh(v^T)\tanh(v) \\ &-k_p \gamma \tanh(\eta_e^T)\tanh(\eta_e) \\ &+k_d \gamma \tanh(v^T)\tanh(v) \\ &+\gamma(k_p - k_d)\tanh(\eta_e^T)\tanh(v) \\ &+\gamma \dot{\eta}_e^T \cosh^{-2}(\eta_e)\dot{\eta}_e \\ &+\cosh^{-2}(v) A \tanh(v)\gamma \dot{\eta}_e^T \\ &-\cosh^{-2}(v)\dot{\eta}_e \dot{\eta}_e^T \gamma B\end{aligned} \tag{43}$$

To achieve boundedness of the system, the last equation must be in the following form:

$$\begin{aligned}\dot{V_{BF_1}} \leq &-\frac{k_{dm}a_m}{b_m}\|\tanh(v)\|^2 \\ &-k_{pm}\gamma\|\tanh(\eta_e)\|^2 \\ &+\gamma\|\dot{\eta}_e\|^2\|\cosh^{-2}(\eta_e)\| \\ &-b_m\gamma\|\dot{\eta}_e\|^2\|\cosh^{-2}(\eta_e)\|^2 \\ &+a_M\gamma\|\dot{\eta}_e\|\|\cosh^{-2}(v)\|\|\tanh(v)\| \\ &+k_{DM}\gamma\|\tanh(v)\|^2\end{aligned} \tag{44}$$

Consider the following inequalities:

$$\begin{aligned}\gamma\|\dot{\eta}_e\|^2\|\cosh^{-2}(\eta_e)\| &\leq \gamma\|\dot{\eta}_e\|^2 \\ a_M\gamma\|\dot{\eta}_e\|\|\cosh^{-2}(v)\|\|\tanh(v)\| & \\ &\leq a_M\gamma\|\dot{\eta}_e\|\tanh(v)\|\end{aligned} \tag{45}$$

The inequality Eq. (44) can be rewritten in the following form:

$$\begin{aligned}\dot{V_{BF_1}} \leq &-\frac{k_{dm}a_m}{b_m}\|\tanh(v)\|^2 \\ &-k_{pm}\gamma\|\tanh(\eta_e)\|^2 + \gamma\|\dot{\eta}_e\|^2 \\ &-b_m\gamma\|\dot{\eta}_e\|^2\|\cosh^{-2}(\eta_e)\|^2 \\ &+a_M\gamma\|\dot{\eta}_e\|\|\tanh(v)\| \\ &+k_{DM}\gamma\|\tanh(v)\|^2\end{aligned} \tag{46}$$

where k_{dm}, k_{pm}, a_m, b_m are the minimum values of the vectors k_d, k_p, A, B, respectively, and k_{DM}, k_{PM}, A_M, B_M are the maximum values of these vectors. In order to simplify the stability study of the above equation, we assume that both minimum and maximum values are equal. So, the expression (44) can be rewritten in the following matrix form:

$$
\begin{aligned}
V_{BF_1} \leq & -\frac{1}{2}\begin{bmatrix}\|\tanh(\eta_e)\| \\ \|\tanh(v)\|\end{bmatrix}^T \overbrace{\begin{bmatrix} 2\gamma k_p & -\gamma(k_p+k_d) \\ -\gamma(k_p+k_d) & \frac{2k_d A}{3B}\end{bmatrix}}^{Q_1} \begin{bmatrix}\|\tanh(\eta_e)\| \\ \|\tanh(v)\|\end{bmatrix}^T - \begin{bmatrix}\|\dot{\eta}_e\| \\ \|\tanh(v)\|\end{bmatrix}^T \\
& \overbrace{\begin{bmatrix} \frac{2B\cosh^{-2}(v)}{3} & -\frac{\gamma A}{2} \\ -\frac{\gamma A}{2} & \frac{2k_d A}{3B}\end{bmatrix}}^{Q_2} \begin{bmatrix}\|\dot{\eta}_e\| \\ \|\tanh(v)\|\end{bmatrix}^T \\
& - \overbrace{\left[\frac{k_d A}{3B} - \gamma k_d\right]}^{\beta_1} \|\tanh(v)\|^2 \\
& - \overbrace{\left(\frac{2B\cosh^{-2}(v)}{3} - \gamma\right)}^{\beta_2} \|\eta_2\|^2
\end{aligned}
\tag{47}
$$

Now, we define the parameters k_d, k_p, A, B so as to ensure that the terms $Q_1, Q_2, \beta_1, \beta_2$ are positive definite. To achieve that, the inequality of linear matrix method is applied. For $Q_1 > 0$, this is equivalent to

$$
\frac{2k_d A}{3B} > 0 \; and \; 2\gamma k_p - \frac{3B\gamma^2(k_d+k_p)}{2k_d A} > 0 \tag{48}
$$

For $Q_2 > 0$, this is equivalent to

$$
\frac{k_d A}{3B} > 0 \; and \; \frac{\gamma B\cosh^{-2}()v}{3} - \frac{3B\gamma^2 A^2}{4k_d A} > 0 \tag{49}
$$

Equation (48) gives:

$$
\frac{k_p}{(k_d+k_p)^2} > \frac{3B\gamma}{4k_d A} \tag{50}
$$

http://dx.doi.org/10.5772/62442

and the Eq. (49) gives:

$$\frac{\gamma A}{4k_d} < \frac{1}{9} \tag{51}$$

since $\cosh^{-2}(v) \leq -1$. For $\beta_1 > 0$, this is equivalent to

$$\frac{k_d A}{3B} - \gamma k_d \geq 0 \tag{52}$$

This gives:

$$\gamma = \frac{A}{3B} \tag{53}$$

For $\beta_2 > 0$, this is equivalent to

$$\frac{2B}{3} \geq 1 \tag{54}$$

Since $\cosh^{-2}(v) \leq 1$, one can write

$$B = \frac{3}{2} \tag{55}$$

substituting Eqs. (53)–(55), yields,

$$A = 4.5\gamma \tag{56}$$

Substituting Eqs. (56) and (55) inlo (50) yields

$$\frac{k_p}{(k_d + k_p)^2} > \frac{9}{36k_d} \tag{57}$$

Using Eq. (19), we can define the parameter k_d as follows:

$$k_d = \|\frac{T}{m}\| - k_g - kp \tag{58}$$

and

$$(k_d + k_p)^2 = (\|\frac{T}{m}\| - k_g)^2 \tag{59}$$

Substituting Eqs. (58) and (59) into Eq. (57), we obtain the second inequality expression

$$36k_p(\|\frac{T}{m}\| - k_g - k_p) > (\|\frac{T}{m}\| - k_g)^2 \tag{60}$$

By solving this inequality, we obtain the value of parameter k_p

$$k_p > \frac{\|\frac{T}{m}\| - k_g}{2} \tag{61}$$

and hence, the value of parameter k_d can calculated as

$$k_d = \|\frac{T}{m}\| - k_g - kp \tag{62}$$

Finally, substituting Eq. (56) into Eq. (51) yields

$$\gamma < 0.31\sqrt{k_d} \tag{63}$$

Now, we calculate the total thrust and the desired orientation. As it is presented previous, the control vector $U = \frac{R^d e_3 T}{m}$ can be written in the standard form as

$$\frac{T^2}{m^2} = U^T U \tag{64}$$

which is equivalent to

$$T = \sqrt{u_x^2 + u_y^2 + u_z^2} \tag{65}$$

Using Eq. (17), the vector $R^d e_3$ can be calculated as

$$R^d e_3 = \begin{pmatrix} \cos(\phi^d)\sin(\theta^d)\cos(\psi^d)+\sin(\phi^d)\sin(\psi^d) \\ \cos(\phi^d)\sin(\theta^d)\sin(\psi^d)-\sin(\phi^d)\cos(\psi^d) \\ \cos(\phi^d)\cos(\theta^d) \end{pmatrix}$$

$$\begin{pmatrix} \frac{u_x m}{T} \\ \frac{u_y m}{T} \\ \frac{u_z m}{T} \end{pmatrix} \tag{66}$$

This leads to obtain the values of φ^d, θ^d

$$\phi^d = \sin^{-1}(\frac{u_1 m \sin(\psi^d) - u_2 m \cos(\psi^d)}{T}) \tag{67}$$

and

$$\theta^d = \tan^{-1}(\frac{u_x \cos(\psi^d) + u_2 \sin(\psi^d)}{u_z}) \tag{68}$$

$$U = \begin{bmatrix} u_x \\ u_y \\ u_z \end{bmatrix} \tag{69}$$

Author details

Yassine Jmili[1*], Nuradeen Fethalla[1], Jawhar Ghomam[1,2] and Maarouf Saad[1]

*Address all correspondence to: nour32004@yahoo.com

1 Department of Electrical Engineering, Ecole de Technologie Superieure, Montreal, Canada

2 Department of Electrical Engineering, CEM-Lab, National School of Engineering of Sfax, Tunisia

References

[1] T. Hamel, R. Mahony, R. Lozano, and J. Ostrowski. Dynamic modelling and configuration stabilization for an x4-flyer. In Proc. of the 15th Triennial World Congress of the International Federation of Automatic Control, Barcelona, 21–26 July 2002.

[2] J.M. Pflimlin, P. Soueres, and T. Hamel. Position control of a ducted fan VTOL UAV in crosswind. *International Journal of Control*, 80(5):666–683, 2007.

[3] M.-D. Hua, T. Hamel, P. Morin, and C. Samson. A control approach for thrust-propelled underactuated vehicles and its application to VTOL drones. *IEEE Transactions on Automatic Control*, 54(8):1837–1853, 2009.

[4] S. Bertrand, T. Hamel and P-L. Helene (2009). Stabilization of Scale Model Vertical Take-off and Landing Vehicles without Velocity Measurements, Aerial Vehicles, Thanh Mung Lam (Ed.), InTech, DOI: 10.5772/6469. Available from: http://www.intechopen.com/books/aerial_vehicles/stabilization_of_scale_model_vertical_take-off_and_landing_vehicles_without_velocity_measurements.

[5] T. Cheviron, T. Hamel, R. Mahony, and G. Baldwin. Robust nonlinear fusion of inertial and visual data for position, velocity and attitude estimation of UAV. In *2007 IEEE International Conference on Robotics and Automation*, pp. 2010–2016, April 2007.

[6] E. Rondon, S. Salazar, J. Escareno, and R. Lozano. *Vision-based position control of a two rotor VTOL mini UAV, John Wiley & Sons*, Inc, pages 191–208, 2013.

[7] K.D. Do, Z.P. Jiang, and J. Pan. On global tracking control of a VTOL aircraft without velocity measurements. *IEEE Transactions on Automatic Control*, 48(12):2212–2217, December 2003.

[8] M. Boutayeb, E. Richard, H. Rafaralahy, H. Souley Ali, and G. Zaloylo. A simple time-varying observer for speed estimation of UAV. In *17th IFAC World Congress*, page CDROM, Seoul, South Korea, July 2008.

[9] K. Benzemrane, G.L. Santosuosso, and G. Damm. Unmanned aerial vehicle speed estimation via nonlinear adaptive observers. In *American Control Conference, 2007. ACC'07*, pp. 985–990, July 2007.

[10] T. Madani and A. Benallegue. Sliding mode observer and backstepping control for a quadrotor unmanned aerial vehicles. In *American Control Conference, 2007. ACC '07*, pp. 5887–5892, July 2007.

[11] B. Landry. Planning and control for quadrotor flight through cluttered environments. Master's thesis, Massachusetts Institute of Technology, Department of Electrical Engineering and Computer Science, 2014.

[12] L. Luque-Vega, B. Castillo-Toledo, and A.G. Loukianov. Robust block second order sliding mode control for a quadrotor. *Journal of the Franklin Institute*, 349(2):719–739,

2012. Advances in Guidance and Control of Aerospace Vehicles using Sliding Mode Control and Observation Techniques.

[13] K.P. Tee, S.S. Ge, and E.H. Tay. Barrier Lyapunov functions for the control of output-constrained nonlinear systems. *Automatica*, 45(4):918–927, 2009.

[14] Eduardo D. Sontag, Mathematical Control Theory: Deterministic Finite Dimensional Systems. Second Edition, Springer, New York, 1998.

[15] M.S. de Queiroz, D.M. Dawson, S.P. Nagarkatti, F. Zhang, "*Lyapunov-based control of mechanical systems*", Boston, MA, USA: Birkhauser, 2000.

[16] S. Bertrand. *Commande de drone miniature a voilure tournante*. Ph.D. thesis, Universite de Nice Sophia Antipolis, 2007.

Section 3

Miscellaneous Topics

Chapter 8

Studies on Effects of Aircraft Noise on Behavior of Rats, Their Plasma Norepinephrine Levels and Cell Morphology of the Temporal Lobe

Guo-qing Di

Additional information is available at the end of the chapter

http://dx.doi.org/10.5772/62586

Abstract

To study the physiological effects of airport noise exposure on organisms, Sprague-Dawley (SD) rats were exposed in soundproof chambers to previously recorded aircraft-related noise for 65 d. As a comparison, unexposed control rats were also used. According to aircraft flight schedules, aircraft noise was replayed and its weighted equivalent continuous perceived noise levels (L_{WECPN}) were adjusted to 75 and 80 dB for the two experimental groups. Rat behaviors were observed through an open field test and the concentrations of plasma norepinephrine (NE) were tested by high-performance liquid chromatography-fluorimetric detection (HPLC-FLD). The morphologies of neurons and synapses in the temporal lobe were also examined by transmission electron microscopy (TEM). Our results indicated that SD rats of experiment group exposed to airport noise of 80 dB had significantly lower line crossing number ($P < 0.05$) and significantly longer center area duration ($P < 0.05$) compared with that of control group. After 29 d of airport noise exposure, the concentrations of plasma NE of experiment group were significantly higher than that of control group ($P < 0.05$). It was determined that the neuron and synapsis of the temporal lobe of experiment group exposed to 80 dB for 65 d showed signs of damage. In conclusion, exposing rats to long-term aircraft noise affects their behaviors, plasma NE levels, and cell morphology of the temporal lobe. Of course, the differences in the hearing sensitivity to different sound frequencies and circadian rhythms between rats and humans can bring variances in physiological effects under the same noise exposure. Therefore, if this study results are applied into humans, it should be further confirmed.

Keywords: Aircraft noise, open field test, norepinephrine, neuron, synapse

1. Introduction

Studies have shown that aircraft noise has a great impact on the health status of populations residing in areas near air traffic, particularly citing cardiovascular diseases and the use of sleep and cardiovascular medications [1, 2]. In addition, researchers have concluded that there is a dose-response relationship between aircraft noise levels and blood pressure of the residents of an area near a military aircraft center [3].

However, most knowledge of the physiological effects of noise is generally obtained through animal experiment. To begin with, the open field test (OFT) is commonly used as a mechanism to assess the neurobehavioral effects of noise. Katz *et al.* [4] showed that white noise of 95 dB increased motor behaviors of rats in OFT and decreased their defecation after 1 h of acute stress. Food intake in OFT is reduced when rats are exposed to white noise of 95 dB, while their defecation increases [5]. In addition, the levels of neurotransmitters or hormones in plasma or brain may reflect the neurobiological effects of noise. Typically, the concentration of norepinephrine (NE) in the brain and cochlea of rats decreases after acute noise stress [6, 7]. Corticosterone levels in mice plasma have been shown to be significantly increased when exposed to acute noise for 10 min [8]. Male rats exposed to broadband white noise of 100 dB have been shown to have a significantly increased level of NE in the brain and corticosterone in plasma [9]. Furthermore, observing the morphological changes in neuronal cells can assess the effects of noise. Outer hair cell apoptosis has been observed in the cochlea of chinchilla after intense impulse noise exposure [10]. In addition, continuous noise stress has been shown to affect both degeneration of epithelial cells and apoptosis of stromal cells in the brain of pig [11]. We have not found any studies indicating that noise exposure induces morphological damage in the temporal lobe, the lobe related to perception and memory [12].

None of the aforementioned studies studied airport noise. Broadband white noise was diffusely applied in previous experimental studies to examine the physiological effects of noise. However, it is important to note that white noise in normal environment is almost non-existent. In China, the aircraft-related weighted equivalent continuous perceived noise level (L_{WECPN}) in many residential areas around airports overstepped the 75 dB limit stipulated by the "Standard of Aircraft Noise for Environment around Airport" policy. In this study, we thus sampled actual aircraft noise and played it back to laboratory rats. We then systematically studied their behaviors, plasma NE levels, and cell morphology of the temporal lobe.

Of course, there are some differences in the hearing sensitivity to different sound frequencies and circadian rhythms between rats and humans, which can bring variances in physiological effects under the same noise exposure. Therefore, if this study results are applied into humans, it should be further confirmed.

2. Materials and methods

When airplanes took off or landed at Xiaoshan International Airport (Hangzhou, China), aircraft noises on the roof of a residential building standing 100 m from the edge of airport

http://dx.doi.org/10.5772/62586

were sampled using an LDS four-channel dynamic signal analyzer (Photon II, Royston, England). Based on the 24-h flight schedule of airport and airplane type, aircraft noises were played through a non-directional dodecahedron sound source (Nor270, Norsonic, Lierskogen, Norway) and the intensity of noise was adjusted by a power amplifier (Nor280, Norsonic, Lierskogen, Norway). The sound absorber and insulation device were optimally assembled so that the L_{WECPN} values of the experimental groups I (EG-I) and II (EG-II) were (75 ± 1.0) dB (L_{Aeq} = 65.3 dB) and (80 ± 1.0) dB (L_{Aeq} =70.3 dB), respectively. In addition, the laboratory was customized to better control acoustics, as the doors were sound-proofed and the vents were installed with mufflers so that background noise was no more than 40 dBA, the highest sound intensity heard by our control group (CG). We measured the intensity of noise exposure with a sound level meter (AWA6291, Hangzhou, China), which was sound-calibrated by a loud-speaker before measurement.

Fifty male Sprague-Dawley rats (6 weeks old, weighing (150 ± 20) g) were purchased from the Experimental Animal Center of Zhejiang University, and were randomly divided into three groups: CG (n = 10), EG-I (n = 20), and EG-II (n = 20). Rats were housed five per cage and maintained in temperature- controlled ((21 ± 3) °C) rooms with cycles of 12 h of light and 12 h of dark (light on at 8:00 a.m. daily), and allowed free access to water and food. Rats were marked on their fur with picric acid to distinguish individuals. Before experiments were carried out, SD rats were bred for 3 days to adapt to the laboratory environment. After this, SD rats in groups EG-I and EG-II were exposed to aircraft noises, while the CG was not exposed. OFT and blood collection for neurotransmitter determination were carried out at 17:00 on Days 1, 8, 15, 22, 29, and 36 after noise exposure (blood collection excluded on Day 36). All data including OFT and neurotransmitter determination were collected on the same rats in CG (n = 5), EG-I (n = 10), and EG-II (n = 10). For investigating long-term effects of airport noise, after 65 d of continuous (excepting time for OFT/blood collection) noise exposure, four rats each were randomly selected from CG and EG-II, respectively, for additional neuronal morphology studies. Animal breeding and experiments were performed in line with the "Quality Management Approach to Laboratory Animals," and all efforts were made to minimize the number of animals used and their suffering.

The size of open field was 100 × 100 × 50 cm, and its bottom divided into 25 grids (20 × 20 cm) by white lines. We termed the nine grids located in the center of the open field as "center area." The open field was located in a 2.0 × 2.0 m audiometric cabin and lit by a 15-watt red lamp for background lighting. We handled the rats by the base of their tails, carried them to the center of the open field, and allowed them to explore the apparatus for 5 min. The behaviors of the rats were tracked and recorded by the camera fixed above the apparatus. The behaviors measured included line crossing number and center area duration.

In order to test the concentration of neurotransmitters in each group, venous blood (1.0 ml) was sampled from the orbital vein. Blood was transferred into 1.5 ml boil-proof microtubes (Axygen, United States) and was kept quiescence for 10 min, and then centrifuged at 4000 r/min for 10 min at 4°C. Next, 200 μl of supernatant was extracted from each sample and 200 μl of 5% perchloric acid was added to it. Their mixture was shaken, left at room temperature for 20 min to fully precipitate the plasma proteins, and centrifuged at 10,000 r/min for 15 min.

Finally, supernatants were filtered with 0.45 μm membrane filters, and high-performance liquid chromatography-fluorimetric detection (HPLC-FLD) was used to measure the concentration of NE. The instrument parameters used for HPLC-FLD were as follows: column, Agilent Zorbax SB-C_{18} column (Agilent, US); mobile phase, methanol-buffer (buffer: 0.07 mol NaH_2PO_4, 10 mmol sodium octanesulfonate, pH 3.5). The gradient procedure of the mobile phase was as follows: at 0 min, 10% methanol and 90% buffer; at 5 min, 10% methanol and 90% buffer; at 30 min, 60% methanol and 40% buffer (1.0 ml/min of flow rate, 20 μl of injection volume, 35.0°C of column temperature, 280 nm of fluorescence excitation wavelength, and 315 nm of emission wavelength). Under these conditions, various substances in plasma were completely separated so that no interference to determination of the targets is experienced.

After being exposed to aircraft noise for 65 d, four rats were randomly selected from the CG and EG-II groups (two per group). We then examined the neuronal and synaptic morphologies of the temporal lobe by transmission electron microscopy (TEM). Rats were anesthetized by administration of an overdose of sodium pentobarbital and then perfused with glutaraldehyde transcardially. The temporal lobe was localized by digital brain stereotaxic instrument (ZH-LanXing B/S, Huaibei, China) with a soft-type cranial drill. After perfusion for about 1 h, we decapitated rats and stripped the whole brain rapidly fixed in glutaraldehyde. After fixed for 24 h, the temporal lobes were removed, cut into thin slices, and further fixed in glutaraldehyde for 3 d. Based on this, slices were washed using PBS, fixed using 1% osmium tetroxide, stained using 2% aqueous solution of uranyl acetate, dehydrated using different concentrations of alcohol and acetone gradient, penetrated and embedded using embedding medium, aggregated in oven, and finally cut into ultra-thin slices stained using 4% uranyl acetate and citrate. Cell structure in these ultra-thin slices was observed by TEM (Philips Tecnai 10, The Netherlands).

3. Results

All data from OFT and HPLC-FLD are expressed as mean ± standard error of the mean (SEM). Differences between means for unpaired samples were tested by one-way analysis of variance (ANOVA) using SPSS Version 16 software. The criterion for significance was $P < 0.05$.

3.1. Open field test

Line crossing number and center area duration were obtained by three individuals working independently, and their mean values were adopted as the final results of OFT.

The line crossing number in OFT is shown in **Figure 1a**, with no significant difference between CG and EG-I over the duration of the experiment. Nevertheless, the line crossing number of EG-II after 8 d of airport noise exposure was significantly less than that of CG ($P < 0.05$). From **Figure 1a**, two conclusions can be drawn: long-term exposure to aircraft noise below L_{WECPN} of 75 dB has no significant impact on the line crossing number of rats, while L_{WECPN} of aircraft noise reaching 80 dB is likely to have an impact on line crossing number in rats.

The result of center area duration in OFT showed that the center area durations of CG and EG-I in OFT are almost unchanged, but the center area duration of EG-II is significantly longer than that of CG ($P < 0.05$) after the 8 d of noise exposure (**Figure 1b**). On other days, center area duration among the three groups showed no significant difference ($P > 0.05$).

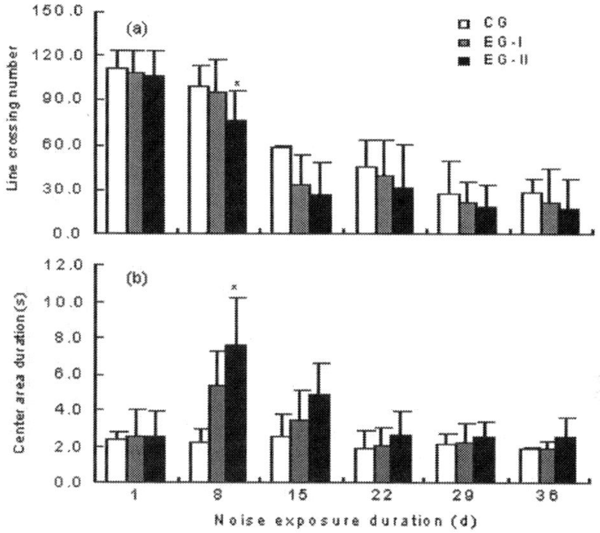

Figure 1. Line crossing number (a) and center area duration (b) in OFT.

There were ten rats in each of EG-I and EG-II and five rats in CG. Data are expressed as mean ± SEM. *$P < 0.05$, compared with the CG (ANOVA).

3.2. Levels of plasma NE

Figure 2 shows the relationship between the time rats were exposed to different intensities of aircraft noise and the average concentration of plasma NE measured. We found that there was no significant difference in NE levels between EG-I and CG over the period of the noise exposure.

Data are expressed as mean ± SEM. *$P < 0.05$, compared with the CG (ANOVA).

Nevertheless, on the 29th day of noise exposure, the levels of plasma NE between CG and EG-II showed significant difference ($P < 0.05$). By analyzing these results, we found that aircraft noise below L_{WECPN} of 75 dB has no significant impact on the plasma NE of rats. Besides, L_{WECPN} of aircraft noise reaching 80 dB is likely to have a negative effect on NE levels under long-term exposure. Therefore, we have known that the intensity of aircraft noise and the duration of aircraft noise exposure are the controlling factors to the level of NE in plasma of rats.

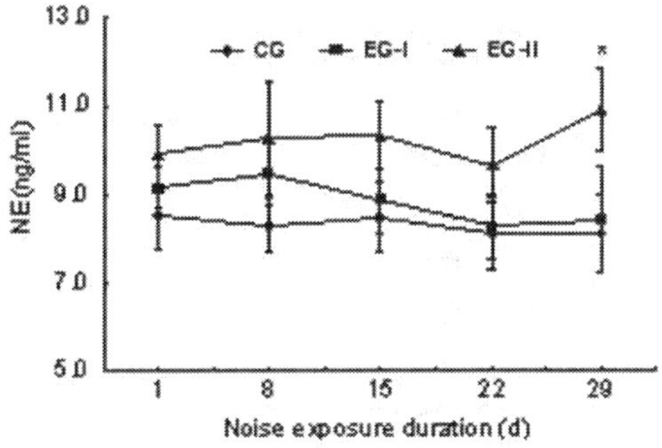

Figure 2. Relationship between NE concentration and noise exposure duration.

3.3. Temporal lobe cell morphology

We observed the neuronal and synaptic morphologies of the temporal lobes from TEM and the representative pictures are shown in **Figure 3**. In our experiments, neuronal and synaptic damages were observed in the temporal lobe of EG-II rats, but no damage was seen in the CG.

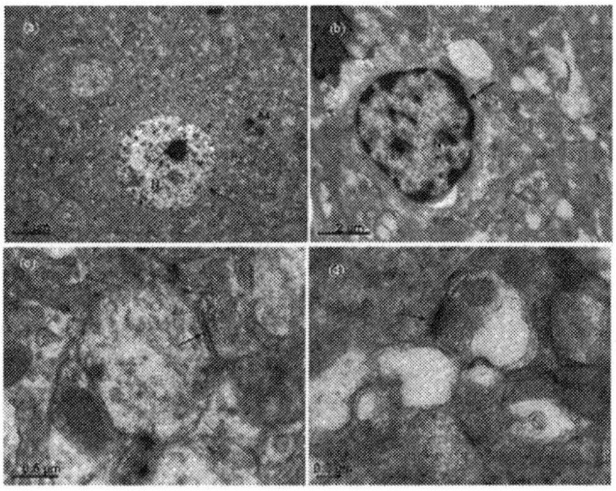

Figure 3. Neuronal and synaptic morphologies of the temporal lobe of rats.

(a) The nuclei (N) in the neurons of the temporal lobe of the CG were oval and their membrane structure was clear, yet mitochondria (M), rough endoplasmic reticula, and other organelles could be seen in cytoplasm, and their distributions were uniform and morphologically normal. (b) The nuclei (N) of neurons in the temporal lobe of EG-II were irregular-shaped, the nuclear membrane was deep-stained and its structure vague, chromatin accumulated along the edge, and cytoplasm was condensed. (c) The synaptic cleft of the temporal lobe area of the CG was clear, and mitochondria and synaptic vesicles.

4. Discussion

4.1. Open field test

First line indent crossing is defined as that rats breach the lattices at the bottom of the open field and the number of line crossing reflects the animals' horizontal mobility, exploration, and anxiety [13]. First and foremost, previous studies have shown that motor behaviors of rats in OFT increased after 1 h of acute stress by white noise of 95 dB [4]. However, in our test, rats of EG-I and EG-II showed no significant difference in line crossing compared with CG after 1 d of noise exposure, which means acute effects of noise exposure were not evident in our experiment. It may be that noise intensity below L_{WECPN} of 80 dB is "moderate" to rats. In addition, after suffering airport noise exposure for 8 d, rats of EG-II showed that the line crossing number decreased, which means horizontal mobility and exploration ability of the rat decreased and some anxiety appeared [13]. However, Pan *et al.* found that two weeks' noise stress (2 h/d, 85 dB) increased square crossing and vertical movement, which is in contrast to our results [14]. There are a number of possible explanations. First, the intensity of noise is an important factor. Second, the duration of daily noise exposure cannot be ignored because, in our test aircraft, noise was exposed throughout the day. Third, the neurobehavioral effect of continuous white noise is different from that of intermittent aircraft noise [15]. Last but not least, from **Figure 1a**, we also know that line crossing of rats was not significantly different between CG and EG-II, except on the 8th day. The reason is that the behaviors of rats manifest itself differently depending on the duration of stress [16]. Conrad *et al.* have also shown that behaviors of rats turned from an excited state to inhibitory state under prolonged stress [17]. Another reason may be that the mechanisms of resistance and/or adaptability are generated after longer term noise stress.

In our study, center area duration is defined as latency of the rats before leaving the center area, which is measured by anxiety-like behavior. As is known, high center area duration indicates high anxiety levels [12]. Thus, the emotions of rats in EG-II altered to anxiety after 8 d of noise exposure. The results of center area duration on other days suggest that the effect of aircraft noise on anxiety is not permanent. This conclusion is also in line with the line crossing results.

4.2. Levels of plasma NE

Epinephrine, NE, and cortisol are stress hormones that are used as indicators of body stress upon noise exposure [6, 18, 19]. NE plays an important role as a stress hormone in conducting and adapting to stress [20, 21]. Therefore, plasma NE level of EG-II increased after long-term noise exposure (29 d) in our experiment, possibly due to the cumulative effect of high-intensity noise. Unfortunately, we do not know the plasma NE level of rats after 29 d of noise exposure, because it is difficult to collect blood from more fierce rats.

Plasma NE is primarily secreted by the sympathetic nerve endings of the heart, blood vessels, and adrenal medulla, and is controlled by the sympathetic nervous system [22]. Plasma NE levels reflect the excitability of the peripheral sympathetic system [23, 24]. Due to aircraft noise exposure in our experiments, the levels of NE in rat plasma increased, indicating that aircraft noise stimulates sympathetic excitement of the adrenal medulla system.

Several studies have shown that trait anxiety is significantly associated with increased NE concentration in blood plasma. Individuals with higher plasma NE concentrations also have more severe anxiety symptoms and the concentrations of plasma NE of patients with anxiety disorders are drastically higher than those of healthy individuals [25]. We conclude that high-intensity aircraft noise exposure may similarly induce anxiety symptoms in rats. Further, patients with hypertension have also been found to have higher plasma NE concentrations [26]. In addition, epidemiological investigations have pointed out that the incidences of heart disease and hypertension are directly related to aircraft noise exposure [1, 27]. Our results partly provide pathological evidence supporting this epidemiological research.

4.3. Temporal lobe cell morphology

Previous studies have shown that necrosis and apoptosis of neurons occur when the body is subjected to physical, chemical, or severe pathological stimulation [28]. For this reason, we consider that the long-term noise stress in rats resulted in the lesions in temporal lobe neurons and synapses of EG-II.

First line indent lobe areas are closely related to perception and memory [29]. Therefore, when neurons of the temporal lobe are damaged, a variety of mental disorders are likely to occur, such as cognitive decline, memory reduction, or subjective emotional instability [30]. When synaptic morphology changes, the related functions of the brain and CNS change accordingly, further leading to changes in behavior [31]. The results of our experiments are consistent with changes in rat behavior due to long-term exposure of aircraft noise.

5. Conclusions

In conclusion, exposing rats to long-term aircraft noise affects their behavior, specifically in the form of inhibiting mobility and increasing anxiety. Our data indicate that scrum NE levels of rats increase as a result of aircraft noise exposure. Furthermore, our findings indicate that aircraft noise exposure leads to damage of neuronal and synaptic structures of the temporal lobe in rats. Nevertheless, additional studies are necessary to further investigate the mechanisms involved.

Of course, there are some differences in the hearing sensitivity to different sound frequencies and circadian rhythms between rats and humans, which can bring variances in physiological effects under the same noise exposure. Therefore, it should be further confirmed if these study results are applied into humans.

Author details

Guo-qing Di

Address all correspondence to: dgq@zju.edu.cn

College of Environment & Resource Sciences, Zhejiang University, Hangzhou, China

References

[1] Franssen E.A.M., van Wiechen C.M.A.G., Nagelkerke N.J.D., Lebret E. Aircraft noise around a large international airport and its impact on general health and medication use. Occup. Environ. Med. 2004; 61 (5): 405–413. DOI: 10.1136/oem.2002.005488

[2] Jarup L., Babisch W., Houthuijs D., Pershagen G., Katsouyanni K., Cadum E., Dudley M.L., Savigny P., Seiffert I., Swart W., et al. Hypertension and exposure to noise near airports: the HYENA study. Environ. Health Perspect. 2008; 6 (3): 329–333. DOI: 10.1289/ehp.10775

[3] Matsui T., Uehara T., Miyakita T., Hitamatsu K., Osada Y., Yamamoto Y. The Okinawa study: effects of chronic aircraft noise on blood pressure and some other physiological indices. J. Sound Vib. 2004; 277 (3): 469–470. DOI: 10.1016/j.jsv.2004.03.007

[4] Katz R.J., Roth K.A., Carroll B.J. Acute and chronic stress effects on open field activity in the rat: implications for a model of depression. Neurosci. Biobehav. Rev. 1981; 5 (2): 247–251. DOI: 10.1016/0149-7634(81)90005-1

[5] Krebs H., Macht M., Weyers P., Weijers H.G., Janke W. Effects of stressful noise on eating and non-eating behavior in rats. Appetite. 1996; 26 (2): 193–202. DOI: 10.1006/appe.1996.0015

[6] Okada A., Ariizumi M., Okamoto G. Changes in cerebral norepinephrine induced by vibration or noise stress. Eur. J. Appl. Physiol. 1983; 52 (1): 94–97. DOI: 10.1007/BF00429032

[7] Vicente-Torres M.A., Gil-Loyzaga P. Noise stimulation decreases the concentration of norepinephrine in the rat cochlea. Neurosci. Lett. 1999; 266 (3): 217–219. DOI: 10.1016/S0304-3940(99)00305-5

[8] Vitale G., Arletti R., Sandrini M. Acute noise stress analgesia in relation to 5-HT2 and μ-opioid receptor changes in the frontal cortex of young mice. Life Sci. 2005; 77 (20): 2500–2513. DOI: 10.1016/j.lfs.2005.01.031

[9] Samson J., Sheeladevi R., Ravindran R., Senthilvelan M. Stress response in rat brain after different durations of noise exposure. Neurosci. Res. 2007; 57 (1): 143–147. DOI: 10.1016/j.neures.2006.09.019

[10] Hu B.H., Henderson D., Nicotera T.M. Involvement of apoptosis in progression of cochlear lesion following exposure to intense noise. Hear. Res. 2002; 166 (1–2): 62–71. DOI: 10.1016/S0378-5955(02)00286-1

[11] Akdogan O., Selcuk A., Take G., Erdoğan D., Dere H. Continuous or intermittent noise exposure, does it cause vestibular damage? An experimental study. Auris Nasus Larynx. 2009; 36 (1): 2–6. DOI: 10.1016/j.anl.2008.03.003

[12] Suzuki W.A., Baxter M.G. Memory, perception, and the medial temporal lobe: a synthesis of opinions. Neuron. 2009; 61 (5): 678–679. DOI: 10.1016/j.neuron.2009.02.009

[13] Walsh R.N., Cummins R.A. The open-field test: a critical review. Psychol. Bull. 1976; 83 (3): 482–504. DOI: 10.1037/0033-2909.83.3.482

[14] Pan F., Lu C.Y., Song J., Jing H., Li Q., Yu H.L., Chen X.Y. Short communication: different duration of crowding and noise exposure effects on exploratory behavior, cellular immunity and HSP70 expression in rats. Stress Health. 2006; 22 (4): 257–262. DOI: 10.1002/smi.1103

[15] Hou G.L. An experimental study on the damaging of non-steady state noise on free cardiac effect radical. Chin. J. Appl. Psychol. 2002; 8 (4): 47–50.

[16] Silveira P.P., Xavier M.H., Souza F.H., Manoli L.P., Rosat R.M., Ferreira M.B., Dalmaz C. Interaction between repeated restraint stress and concomitant midazolam administration on sweet food ingestion in rats. Braz. J. Med. Biol. Res. 2000; 33 (11): 1343–1350. DOI: 10.1590/S0100-879X2000001100013

[17] Conrad C.D., Magariños A.M., LeDoux J.E., McEwen B.S. Repeated restraint stress facilitates fear conditioning independently of causing hippocampal CA3 dendritic atrophy. Behav. Neurosci. 1999; 113 (5): 902–913. DOI: 10.1037/ 0735-7044.113.5.902

[18] Vaernes R., Ursin H., Darragh A., Lambe R. Endocrine response patterns and psychological correlates. J. Psychosom. Res. 1982; 26 (2): 123–131. DOI: 10.1016/0022-3999 (82)90030-7

[19] Spreng M. Possible health effects of noise induced cortisol increase. Noise Health. 2000; 2 (7): 59–63.

[20] Goldstein D.S. Plasma norepinephrine as an indicator of sympathetic neural activity in clinical cardiology. Am. J. Cardiol. 1981; 48 (6): 1147–1154. DOI: 10.1016/0002-9149(81) 90333-7

[21] Finlay J.M., Zigmond M.J., Abercrombie E.D. Increased dopamine and norepinephrine release in medial prefrontal cortex induced by acute and chronic stress: effects of diazepam. Neuroscience. 1995; 64 (3): 619–628. DOI: 10.1016/0306-4522(94)00331-X

[22] Esler M., Jennings G., Lambert G., Meredith I., Horne M., Eisenhofer G. Overflow of catecholamine neurotransmitters to the circulation: source, fate, and functions. Physiol. Rev. 1990; 70 (4): 963–985.

[23] Lake C.R., Gullner H.G., Polinsky R.J., Ebert M.H., Ziegler M.G., Bartter F.C. Essential hypertension: central and peripheral norepinephrine. Science. 1981; 211 (4485): 955–957. DOI: 10.1126/science.7466370

[24] Raskind M.A., Peskind E.R., Halter J.B., Jimerson D.C. Norepinephrine and MHPG levels in CSF and plasma in Alzheimer's disease. Arch. Gen. Psychiatry. 1984; 41 (4): 343–346.

[25] Yasunari K., Matsui T., Maeda K., Nakamura M., Watanabe T., Kiriike N. Anxiety-induced plasma norepinephrine augmentation increases reactive oxygen species formation by monocytes in essential hypertension. Am. J. Hypertens. 2006; 19 (6): 573–578. DOI: 10.1016/j.amjhyper.2005. 10.027

[26] Makino S., Iwata M., Fujiwara M., Ike S., Tateyama H. A case of sleep apnea syndrome manifesting severe hypertension with high plasma norepinephrine levels. Endocr. J. 2006; 53 (3): 363–369. DOI: 10.1507/endocrj.K05-169

[27] Knipschild P. V. Medical effects of aircraft noise: community cardiovascular survey. Int. Arch. Occup. Environ. Health. 1977; 40 (3): 185–190. DOI: 10.1007/BF01842081

[28] Bonfoco E., Krainc D., Ankarcrona M., Nicotera P., Lipton S.A. Apoptosis and necrosis: two distinct events induced, respectively, by mild and intense insults with N-methyl-D-aspartate or nitric oxide/superoxide in cortical cell cultures. PNAS. 1995; 92 (16): 7162–7166. DOI: 10.1073/ pnas.92.16.7162

[29] Baxter M.G. Involvement of medial temporal lobe structures in memory and perception. Neuron. 2009; 61 (5): 667–677. DOI: 10.1016/j.neuron.2009.02.007

[30] Hugdahl K., Løberg E.M., Nygård M. Left temporal lobe structural and functional abnormality underlying auditory hallucinations in schizophrenia. Front. Neurosci. 2009; 3 (1): 34–45. DOI: 10.3389/neuro.01.001.2009

[31] de Bartolomeis A., Fiore G. Postsynaptic density scaffolding proteins at excitatory synapse and disorders of synaptic plasticity: implications for human behavior pathologies. Int. Rev. Neurobiol. 2004; 59: 221–254. DOI: 10.1016/S0074-7742(04)59009-8.

Chapter 9

Numerical Study of Support Interferences on the SOAR Separation Wind Tunnel Test

Alberto Ghiraldo, Sebastien Paris and Ernesto Benini

Additional information is available at the end of the chapter

http://dx.doi.org/10.5772/62241

Abstract

A process of support design for wind tunnel models and the evaluation of interferences effect are described in this chapter. The work was performed at the Von Karman Institute for Fluid Dynamics (VKI; Sint-Genesius-Rode, Belgium), and it was commissioned by the S3-Swiss Space System company. The work concerns the separation wind tunnel test of the Suborbital Aircraft Reusable (SOAR) vehicle from an Airbus commercial plane carrier. The supports are designed for future separation wind tunnel test of the SOAR version V10 in the VKI-S1 wind tunnel. They are designed in scale 1:180 for the test of the SOAR in the presence of the Airbus and in scale 1:80 for the SOAR alone test. Two different shapes of support (circular and elliptic) are tested in each case. First there are the supports designed, then the results of the finite element method (FEM) static structural analysis and vibrational analysis, and finally the result of the computational fluid dynamics (CFD) campaign. The flow and the force interferences caused by the support are investigated by comparing simulations with and without support. The behavior of the two shapes and of the dimensional variations are investigated at an angle of attack between 0° and 15° and at Mach 0.7.

Keywords: separation wind tunnel test, SOAR, support design, support interferences, VKI

1. Introduction

Theoretical aerodynamic study, computational fluid dynamics (CFD) analysis, and wind tunnel tests are the three pillars of every aerodynamic design process. In this paper, the emphasis is on how CFD and theoretical study can be complementary and how to prepare the wind tunnel test.

The present project, commissioned by the S3-Swiss Space System, deals with the development and qualification of three stages to orbit a composite vehicle. The first and second stages, the Airbus and the Suborbital Aircraft Reusable (SOAR) vehicles, respectively, are reusable, whereas the third stage is an expandable booster. All details of the SOAR mission are in Ref. [1] see **Figure 1**. This project requested the design of new supports for the separation wind tunnel test in the VKI facility and the study of the respective interferences.

The S3 SOAR separation from the Airbus A300 is a critical issue during the mission. The aerodynamic coefficients are affected by the proximity of the vehicles. To correctly design the separation, the aerodynamic database must be known with sufficient accuracy. For this reason, an important campaign of wind tunnel testing is necessary.

An earlier wind tunnel campaign had already been performed with the construction of an initial aerodynamic database [2]. To carry out further wind tunnel experiments, an improvement in the knowledge of support design is required. In fact, to design supports for the SOAR, in this configuration, is particularly complicated due to the presence of the Airbus below and behind the vehicle (the A300 empennage also impedes the use of a support from the base). Effectively, the first sting designed produced considerable interferences on the model. Therefore, the purpose of this work is to design supports for the SOAR vehicle in the following two configurations: scale 1:180 for the wind tunnel test of the SOAR in the presence of the Airbus and scale 1:80 for the wind tunnel test of the SOAR alone.

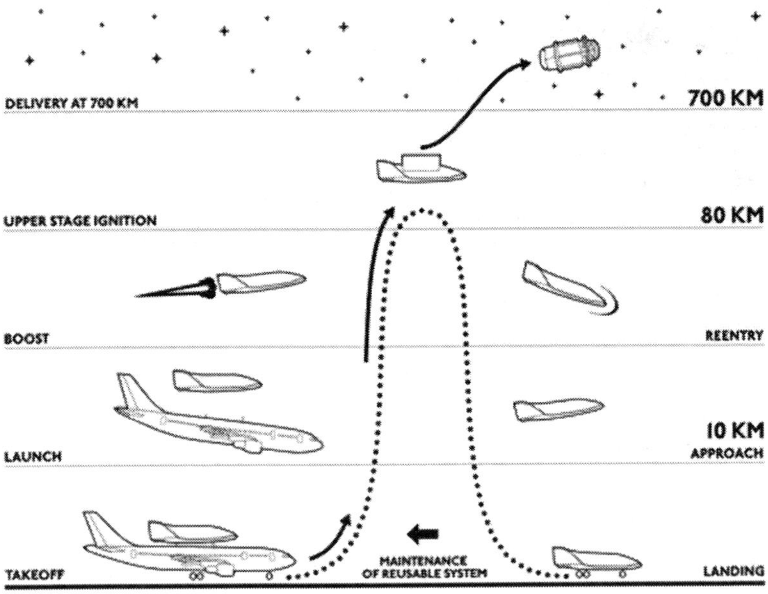

Figure 1. Illustration of the main steps of the mission.

2. Support design

The support in wind tunnel tests is necessary to hold up the model in the test section, but it is also an artificial device that, especially from an aerodynamic point of view, does not exist. From this definition, the following design constraints are immediately derived: to minimize the aerodynamic interferences with the flow ensuring adequate mechanical properties to sustain the model. It is also necessary to consider the wind tunnel test conditions (M=0.7, α_{max}=15°), the model shapes, and the use of steel as material for the supports. Furthermore, the support has to allow the internal passage of cables that interface the 6 degrees of freedom internal balance with the measurement device. Finally, the test section dimensions of the VKI-S1 wind tunnel (0.4×0.36 m) and the need to adopt a new configuration with respect to the previous one are considered. This is because the use of a dorsal strut (see **Figure 2**) did not provide satisfactory results regarding the interferences with the models. Some additional constraints are present in scale 1:180 composite testing: to test the separation with the presence of the Airbus empennage and to consider the nominal relative position of the two vehicles. This is necessary to permit a relative motion of the SOAR over the Airbus.

Once requirements about flow measurement and passage of cables are acquired, the two parameters to be optimized are to minimize the aerodynamic interferences and to give adequate mechanical properties.

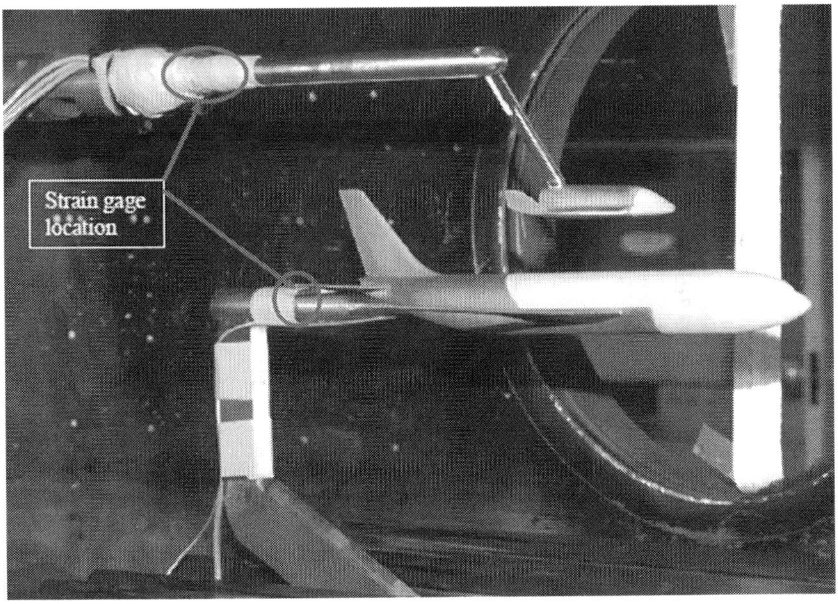

Figure 2. Picture of the old support used in the first separation wind tunnel test.

In fact, to minimize interferences, the sting has to be as long as possible and with the diameter as small as possible. These requirements are not in agreement with the structural properties for which stings should be short and have a large diameter. An important parameter in this dissertation is the "critical sting length" that is "the shortest sting length, which does not change the level of an aerodynamic measurement obtained with longer stings" [3]. The critical sting length is influenced by the Mach number, Reynolds number, boundary layer at the base of the model, sting, and model base diameter. The Reynolds number plays an especially important role. In fact, if the flow is laminar at the base of the model, L_c is as much as 12 to 15 times the base model diameter (D); on the contrary, with turbulent flow, L_c is reduced to 3–5·D.

It is necessary to pay attention that the diameter does not have to modify the typology of the boundary layer at the model base. The minimum diameter allowable, from load considerations, is approximately 0.25 times the model base diameter [4]. For the maximum value, it is necessary to consider that, in the transonic flow, minimum interferences exist with a sting diameter up to 0.4 times the base [5].

In this flow field, the common choice is to use a "straight sting," which is the best solution to reduce the interferences. This is a tube that enters the model at the base. With this choice, all the supporting structure is downstream of the model and it is used with an internal balance system [6].

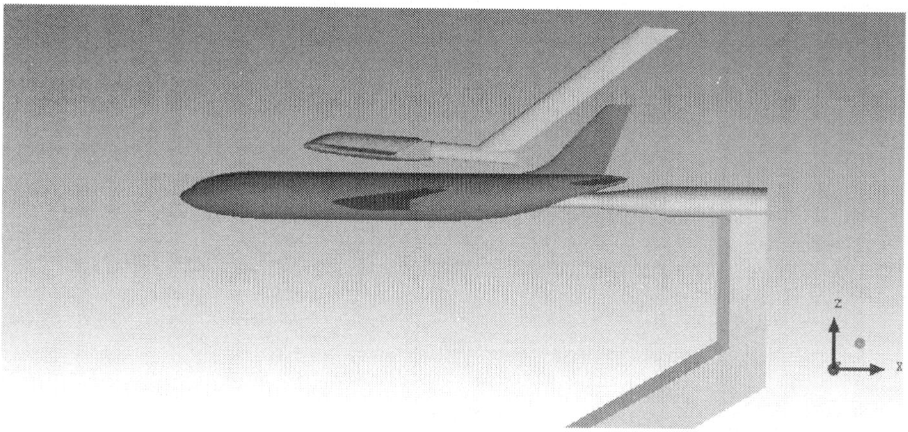

Figure 3. The SOAR (light blue) with the new support (orange) over the Airbus (dark green) in the nominal position; scale 1:180.

The use of a straight sting is possible only for the scale 1:80 model, see **Figure 3** whereas the presence of the Airbus, with its empennage, in the composite configuration (scale 1:180) prevents its usage. see **Figure 4** The only remaining possibility, in this case, is to use a straight sting that enters in the SOAR base (rear surface) but at the other extremity is connected with an inclined bar before the Airbus empennage. The presence of the cables inside the support

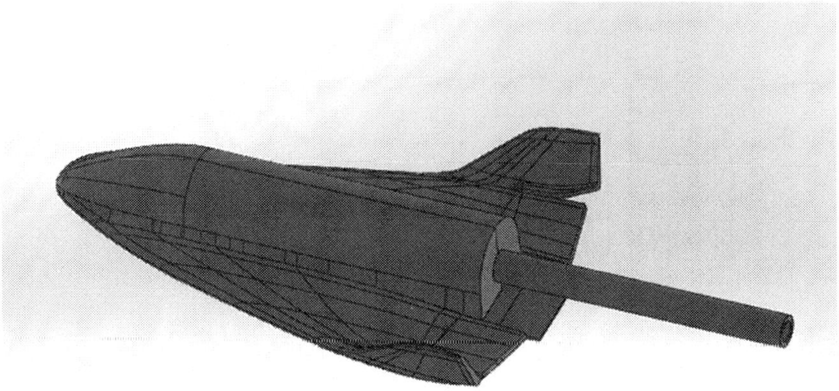

Figure 4. The SOAR with the new circular sting; scale 1:80.

needs a cavity that decreases the stiffness, with the necessity to pay particular attention to the thickness of the walls. From the definition of the critical Reynolds number, it is possible to see that, for the scale 1:80, the flow at the base is definitely turbulent, whereas, in scale 1:180, the transition is around the base. In any case to design a sting 12 to 15 times the length of *D* is not possible due to the constraint of the Airbus empennage, and the longest one possible was designed.

Also, for the diameter, in scale 1:180, it was not possible to respect the rules explained for the presence of the cable inside, and again the smallest one possible was chosen.

The inclined part of the support is clearly a critical point for the interferences and it dramatically breaks the flow in the proximity of the rear portion of the SOAR. For this reason, it made it as less as possible inclined (*B* angle small) with respect to the constraint of the empennage angle (otherwise the support would approach the tail of the Airbus too closely). Another solution, adopted to try to have small interferences, is to use an airfoil shape as section of the inclined bar. In fact, in the transonic flow, a thin airfoil allows for a dramatic reduction of the drag [7]. The aim of this is to choose the thinnest airfoil possible, which is compatible with the structural constraints and the presence of cables inside. The choice was a NACA 0016 cut in the rear part.

Two different shapes of sting sections are evaluated in this work: circular and elliptic. The circular one is the most common and there are many references in the literature [3, 4, 6]. On the contrary, an elliptic sting should reduce the interferences and the drag especially at a high angle of attack [8]. These configurations will be evaluated at every step of the design process, first from a structural point of view and then with a flow analysis using a CFD software. Because there is a lack of literature on the use of the elliptic sting, its use and the possible advantages of this choice are investigated in depth. All support dimensions are in Ref. [9].

3. Structural analysis

To evaluate the reaction of different supports at the external loads present in the wind tunnel, a static structural analysis and a vibrational analysis were performed [finite element method (FEM)].

For the static structural analysis, the contribution of drag, lift, pitching moment, and lateral forces was taken into account. From this study, although it results that the external loads do not involve compromising stresses, strains, and displacements, it is necessary to highlight two critical points. The first one is that the bending angle due to the displacement at the free extremity, where the model will be connected, is often relevant, especially in scale 1:80. In all cases (circular and elliptic sections, scales 1:80 and 1:180), it is necessary to take into account this bending angle, with an accurate measurement, to have the exact inclination of the model. The second interesting consideration concerns the larger stress (Von Mises) concentrations present with the elliptic sting. The worse behavior of the elliptic shape is because it has a worse filling. In fact, to fit the same cables, it is necessary to reduce the thickness of the wall; otherwise the length of the major semi-axis would have been too large. Both problems are manageable, but they require attention.

With the frequency analysis, it was verified that the external excitation during the experiments is far enough from the resonance frequency of the supports. The main vibrational external disturbance for the stings is the frequency of the vortex shedding of the SOAR model. Thus, the utilization of an elliptic section is possible, but it needs more vigilance regarding the designing of the wall thickness with an increase of the overall dimensions.

4. CFD analysis

The CFD simulations represent the most important contributions in this project. In fact, the flow field and the force interferences due to the support were studied by comparing simulations with and without support. To study the alterations of the flow during the wind tunnel test due to the presence of the artificial strut is fundamental for two reasons: to help design it as discretely as possible and to understand the appropriate correction to the wind tunnel results. In particular, the impact of the different shapes of stings (circular and elliptic), the dimensions, the presence of an inclined part (with different inclinations), and the effect of an angle of attack were studied. Every shape of support (circular or elliptic) was tested, with the nominal dimensions, connected to the vehicle, at angles of attack of 0°, 5°, 10°, and 15°. After this, the variation of the dimensions was evaluated at 0°, with the sting longer or shorter by 10%. In addition, simulations with the diameter bigger than 10% were performed but not with a smaller one, which is not possible if the cables are to be bunched inside the sting. A change in the dimensions (10% length and diameter increase) is applied also with the presence of an angle of attack to understand how these two different aspects can interact. The effect of the inclined part (scale 1:180) was studied by simulations with the nominal inclination of the vertical part compared to the straight sting (40.8°) and adding 5°, 10°, and 15° of inclination, respectively.

4.1. Pre-processing operations

All meshing operations were done using Ansys ICEM CFD version 15.0, whereas the selected CFD tool was Metacomp CFD++ version 14.1. The computations were performed on the ClusterVision cluster of the VKI.

The unstructured triangular surface mesh was realized using the autoblock method. Elongation surfaces were inserted near the SOAR trailing edge to delay the boundary layer to tetra mesh transition away from the body. This ensured compromising pressure jumps, and it might appear far from the trailing edge due to element transitions. From this surface, a tetra mesh see **Figure 5** mixed with a Delauny method volume mesh was built between the model and the fluid domain walls. After this, an exponential "prism growing" extrusion method was applied, which is a kind of a level-set technique to grow constant thickness layers around the body (with increasing thickness layer by layer). This prism layer forms the boundary layer thickness and it consists of 25 levels with an expansion ratio of 1.2 and first cell height of 0.5 mm. During growth, a directional smoothing and a smoothing of the first layer were applied. Some pyramid elements were generated to connect the prism column end sides to the tetra mesh. Because the target Mach number was lowered to 0.7, it went below the critical Mach number; therefore, shock refinement became unnecessary for this study. However, very small pockets of supersonic flows were observed with Mach number less than 1.1. The total number of mesh elements is just over 1.4 million.

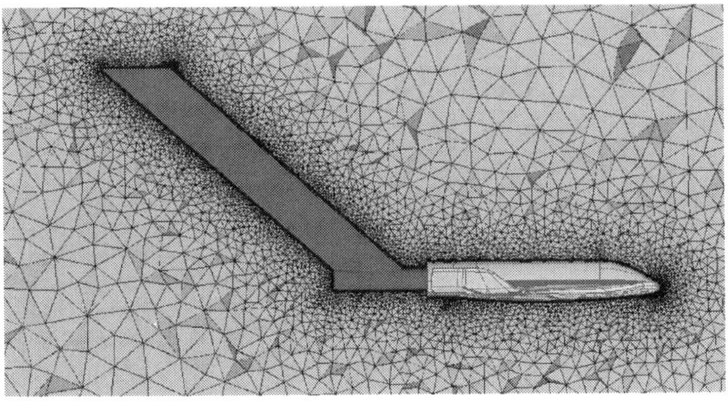

Figure 5. Solid and wire view of the completed volume mesh symmetry plane.

In CFD++, the RANS simulations were performed with the k-ω SST model. All simulations were at M=0.7 and with the wind tunnel set-up [9]. The body-support intersection is quite large; therefore, its effect could not be neglected. The open part would behave as a zero pressure region if not corrected. Hence, the following correction was applied: around the holes of the

SOAR, where the support attaches to the body, the average pressures were computed. Based on these pressures and on the area of the hole, it is possible to introduce a compensating axial force see **Figure 6**.

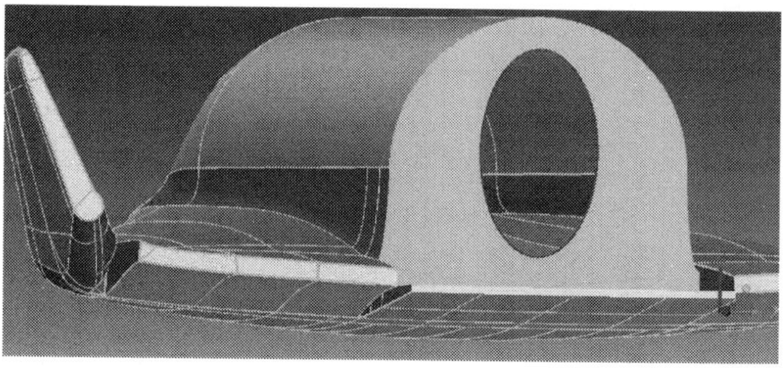

Figure 6. View of the SOAR with the hole for the support. The support correction is based on the average pressure on the red line.

4.2. Result of the CFD analysis

The results were evaluated by looking at the flow topology and at the aerodynamic forces and coefficients database. A summary of the result is presented in the following paragraphs, the complete analysis is in Ref. [9].

4.2.1. Scale 1:180

The origin of the interferences on the model is twofold. The presence of the support blocks the flow in the wake, decreases the recirculation, and increases the pressure that pushes the body forward, reducing the drag. Furthermore, the blockage of the flow in the narrow space between the flap surfaces and the sting generates an extra pressure on the flaps. Because this downward push on the flaps is far from the vehicle's center of gravity (CoG), it causes a considerable increase on the pitching moment. Second, the inclined bar of the support breaks the flow towards the SOAR, especially on the upper surfaces. This generates over pressures on the body and, again, on the flaps responsible for the lift decrease and a contribution in the growing of the moment.

Moreover, the lower surface of the vehicle reveals a higher pressure, but the values are lower than those on the upper part and arranged more homogenously. Then, the higher pressure on the bottom rather contrasts the decrease of the lift but less the increase of the pitching moment.

Now focusing on the effect of different sting shapes, the elliptic one with its more streamlined cross-section causes fewer interferences allowing a better ease of the flow moving around. In particular, this phenomenon is present below the circular sting that, with its larger section, stops the flow more. Furthermore, the circular sting induces a larger vortex on its top.

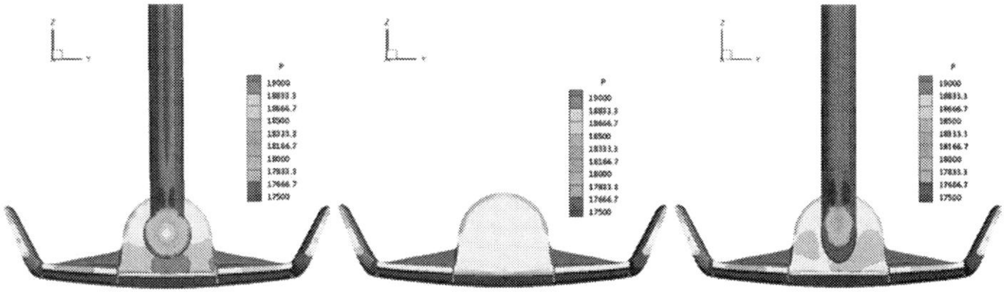

Figure 7. Rear surface view of the SOAR with circular sting on the left, alone in the middle, and with elliptic one on the right; scale 1:180, α=0°.

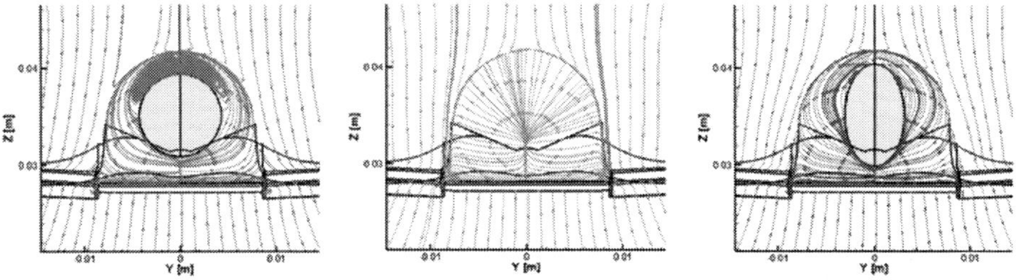

Figure 8. Plot of the y and z velocity components; the SOAR with circular sting on the left, alone in the middle, and with elliptic one on the right; scale 1:180, α=180°.

The impact of these flow alterations on the SOAR coefficients is minimal concerning the drag and the lift, with small angles of attack, for both shapes and it increases at 15°. Raising α the interferences usually growth, less using the elliptic section, that is now clearly advantageous, especially on the lift. On the contrary, for the moment coefficient, the interferences are more important also at small angles presenting a linear increase with the angle of the flow. Even in this case, the usage of an elliptic sting produces less interference. In **Table 1**, there is the average correction on the SOAR alone coefficient, for values of α between 0° and 15°, in scale 1:180. The correction is calculated as the value of the coefficient when the SOAR was alone, minus the value of the same coefficient, only for the SOAR, in a simulation where the support was also present.

Simulations	$\Delta C_{L\,av}$	$\Delta C_{D\,av}$	$\Delta C_{M\,av}$
SOAR with circular sting	0.00590	0.00928	−0.03831
SOAR with elliptic sting	0.00451	0.00703	−0.03425

Table 1. Average correction, respect of the angle of attack, on the coefficients when the support is present compared to the SOAR alone; scale 1:180

To understand the impact of different dimensions on the sting performance, 10% alterations were chosen on the length (longer and shorter) and on the diameter (only larger) see **Figure 9**. It is clear that a shorter sting presents higher interferences with respect to a longer one. In fact, the shorter one induces a higher pressure on the model rear surface and an overpressure on the upper surface and on the flaps. This last fact is due to the greater proximity of the vertical part of the support that breaks the flow over the vehicle.

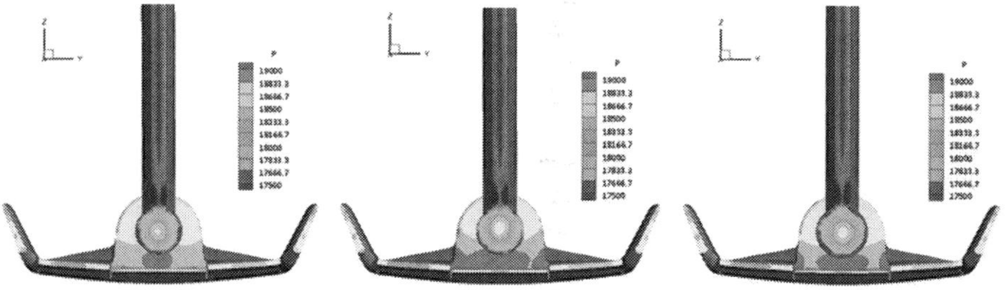

Figure 9. Rear surface view of the SOAR with the sting 10% longer on the left, 10% shorter on the middle, and with the bigger diameter on the right; scale 1:180, α=0°.

It is interesting that a 10% length reduction produces larger changes than a 10% increase. In fact, regarding lift and drag, the effect on the coefficients is smaller when the sting is longer (4–10% interferences see **Figure 11** reduction) but more important when it is shorter (30–40% interferences increases) with respect to the nominal dimensions. This does not happen for the circular sting drag coefficient where it is the opposite. About the pitching up moment, the same tendency is present but with minimal variations, and it is not possible to notice any differences between the two shapes.

An increase of 10% in the diameter produces 17% to 19% of interference growing on the pitching up moment, whereas, on the drag and on the lift, the worsening is small. The larger impact on the pitching moment is explainable with the grater proximity of the sting to the flap.

The variation of the dimension was evaluated also at α=15°, observing that it has a greater impact on the coefficients than what happens at α=0°. The trend for the drag and the lift is almost the same as with α=0°. Interesting is the drastic reduction of pitching moment interference with the longer elliptic sting (144%), whereas this is not present with the circular shape. By increasing the diameter, the coefficient most negatively affected is the pitching moment.

Increasing the inclination of the vertical bar, the differences on the pressure distributions are really appreciable only with 15° of variation (this means that the bar is closer to the vertical axis). It is possible to see an increase of the pressure on the rear surface and on the upper one, especially in the rear part. Otherwise, on the bottom, it is not possible to notice strong differences see **Figure 10**. This, as already clarified, means a decrease of the drag and the lift while the pitching moment grows. The global result is an increase on the interferences, which should be taken into account.

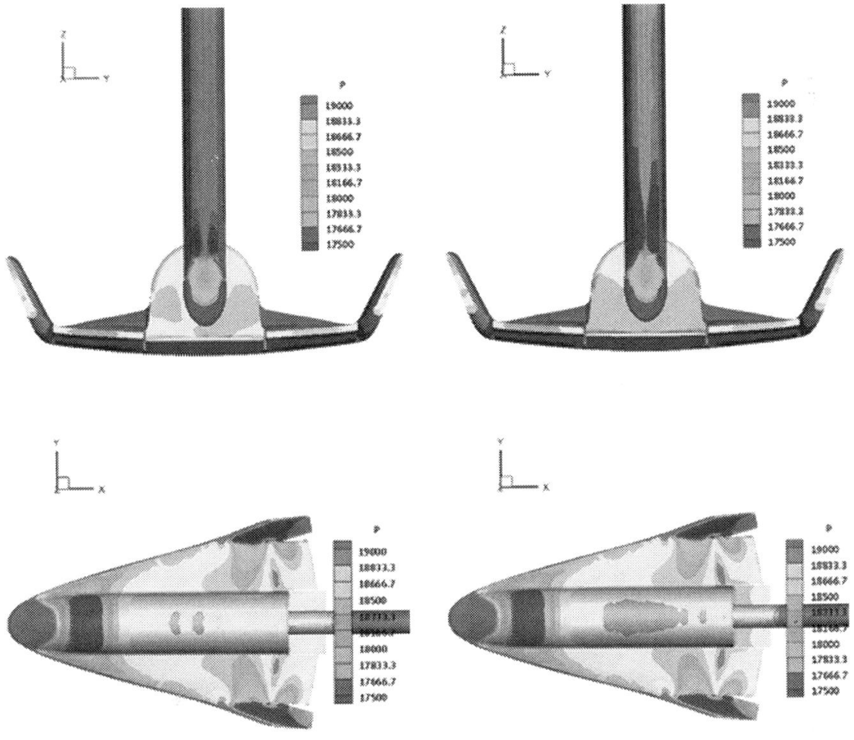

Figure 10. Different views of the SOAR with elliptic support. B=40.8° (left) and B=55.8° (right); scale 1:180, α=0°.

4.2.2. Scale 1:80

As expected, straight stings disturb less the flow than any supports with an inclined part. An increase of pressure on the model rear part due to the presence of the sting is present again but, because there are no large parts sticking out of the wake, the effects are small. The pressure that increases over the vehicle surface is much lower, but an increase on the rear surface of the SOAR is still present. This last phenomenon is more evident with the circular sting than with the elliptic one. Once again, the circular shape induces a larger blockage of the flow below the sting.

On the coefficients too, the straight sting induces the same kind of interferences, that is, the decrease of the drag and the lift and increase of the pitching moment. This time, the deviance with respect to the drag and to the lift from the SOAR alone is actually small see **Table 2**. For the pitching moment, greater interferences persist and, once again, the elliptic sting reacts better. This is due to the presence of the SOAR flaps under the sting that induces a higher pressure again.

Simulations	$\Delta C_{L\ av}$	$\Delta C_{D\ av}$	$\Delta C_{M\ av}$
SOAR with circular sting	0.00229	0.00254	−0.01434
SOAR with elliptic sting	0.00186	0.00217	−0.00857

Table 2. Average correction, respect of the angle of attack, on the coefficients when the support is present compared to the SOAR alone; scale 1:80

The 10% dimensional variations on the straight sting produce minimal differences on the pressure contours. Now a shorter sting causes a lower pressure on the vehicle rear surface with a behavior closer to the SOAR alone. The reasons for this are twofold. First, it should be remembered that turbulent flow at the model rear base predicts a shorter critical sting length (three to five times the base diameter) that is not possible to predict exactly. The sting length in scale 1:80 is 3.4·*D* and the critical sting length should be on the range of 10% variations. Second, with this model of sting, there are not any inclined parts out from the wake of the SOAR. In fact, the disturbances of the support, in scale 1:180, are bigger when the vertical part is closer to the vehicle, that is, with a shorter sting. On the contrary, with only a straight sting, the shorter it is, the lower are the perturbations on the wake, even if the amount of the interferences is always smaller. It is difficult to clarify this trend because in reality, at the end of the sting, there should be another structure usually with a larger diameter. The effect of this bigger diameter component can be similar to what induced an inclined bar, that is, to increase the flow blockage as closer to the vehicle. Now the effect on the coefficients is smaller and they usually remain approximately constant with some exceptions: on the lift, the elliptic longer sting reduces the interferences (29%), whereas the circular longer one increases (16%). On the pitching moment, the longer elliptic sting reduces the interferences by approximately 20%.

Increasing the diameter by 10% produces similar interferences to what happens in scale 1:180. Once again, the greatest deviances are observed for the pitching moment.

The variation of the dimensions at 15° angle of attack has a small impact on the drag and on the lift, whereas, on the pitching moment, there is a drastic reduction of interferences with both shapes. Increasing the diameter causes only a 20% growth of the deviation for the moment coefficient with respect to the SOAR alone.

4.3. Comparison with the previous support

It is useful to remember that this study and the support design rose from the necessity to build new supports after the first wind tunnel testing campaign in the VKI. All results of that research activity are found in Ref. [1].

The old support was attached to the vehicle from the top (dorsal strut) and it was discovered to induce high interferences. That support configuration induced an extra pressure in front of the sting and lower pressure on the side. Furthermore, lower pressures were presented on the back of the SOAR behind the body and on the top of the flaps. These pressure decreases induced large forces due to the big areas it was acting on with the double effect to increasing in the drag and generating a pitching down moment.

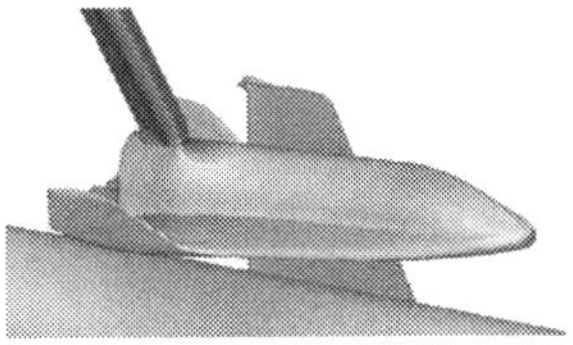

Figure 11. Sting effect on the old support. It is possible to see the red peak of pressure at the base of the sting; scale 1:180.

The previous separation wind tunnel tests were performed in scale 1:180, and the cases at 0 and 10° angle of attack are comparable with the results of this paper.

The use of the posterior supports dramatically reduces the interferences in terms of drag and lift with respect to the dorsal strut. On the contrary, the rear sting causes greater interferences to the pitching moment. Furthermore, the sting effect is opposite: whereas the drag was increased, now it is decreased, and the pitching moment now is higher, whereas before it was smaller. Also, the effect with the angle of attack is different: with the previous support, the amount of the correction was reduced increasing α, whereas the new models make the situation worse with high α. Only for the lift all choices cause decrease in the coefficient respect to the SOAR alone see **Table 3**.

	Old support			New circular support			New elliptic support		
α	ΔC_D	ΔC_L	ΔC_M	ΔC_D	ΔC_L	ΔC_M	ΔC_D	ΔC_L	ΔC_M
0	−0.0328	0.0662	0.0082	0.0045	0.0086	−0.0242	0.0039	0.0078	−0.0218
10	−0.0159	0.0891	0.0059	0.0058	0.0079	−0.0426	0.0046	0.0071	−0.0391

Table 3. Comparison of the performance between the support with circular and elliptic sting and the old one; scale 1:180, α=0°

4.4. Local pressure coefficient corrections

In this paragraph, the coefficient alterations are revised, but now looking at the pressure point by point. The following critical area of the vehicle are selected for this analysis: the rear surface, the body, and the flaps see **Figure 12**.

Figure 12. With different colors, the lines where the pressure are evaluated: left, the SOAR rear surface; middle, the body; right, the flap.

The main goal of the investigations is to find corrective coefficients to correct for the presence of the support with information coming from the pressure values. In fact, during the wind tunnel test, a small transducer will be positioned to measure the pressure at smart locations. With the coefficients found in this analysis, it will be possible to have a value of the local pressure corrected and then close to what happens without the support. Afterwards, it is desirable to apply these coefficients also at corrective laws for the aerodynamics coefficients. The sense of the last point is to have a quick way to obtain an indicative value of the coefficient correction for each case, which can be tested in the wind tunnel. The study is done for the nominal dimension geometry function of the angle of attack.

First, the local pressure values are presented, and after the correction, laws are explained.

4.4.1. Local pressure values

Regarding the rear surface and the upper body surface, what was already analyzed in the previous paragraph was confirmed. The circular sting induces higher pressure in the lower part of the rear surface, with a large peak and slightly lower value in the upper zone see (see **Figure 13**) Increasing the sting length, as expected, the pressure is closer to the values of the SOAR alone (see **Figure 9**). This is present with the same trend for both sting shapes in scale 1:180, not in scale 1:80 where the differences are small.

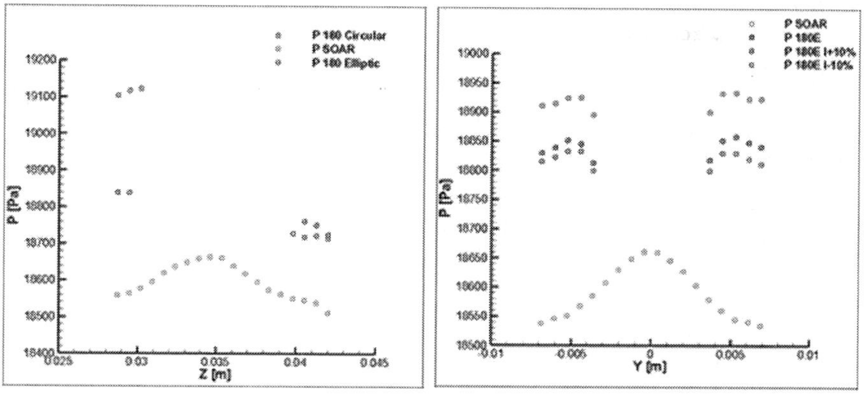

Figure 13. Rear surface pressure distribution plot: on the red line for the elliptic sting with different length (left) and on the blue line for SOAR alone and with different sting shapes (right); scale 1:180, α=0°.

On the body upper surface, the support creates an overpressure on the rear part of the vehicle, more evident in scale 1:180. The pressure is calculated also on the right flap in the longitudinal direction following the black line shown in **Figure 12**. As expected, the longer sting always has a lower correction than the shorter and the elliptic shape is closer to the SOAR alone (see **Figure 14**).

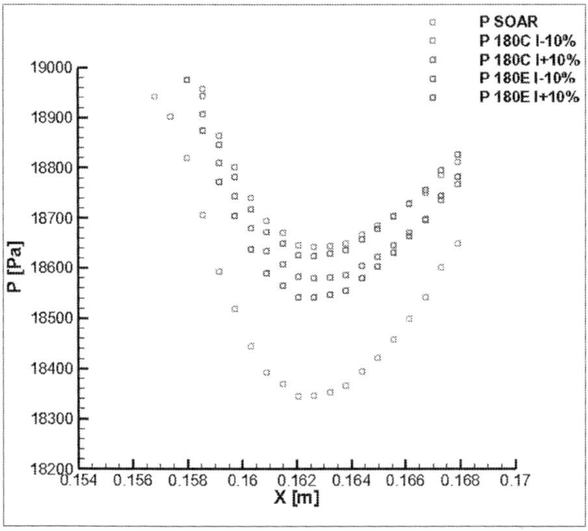

Figure 14. Pressure distribution on the flap for different shapes and geometries; scale 1:180, α=0°.

The impact of higher pressure on the flap is always towards increasing the pitching up moment. It is interesting to notice that the flap is more affected by the sting for α=0° than for α=15° and that the circular sting produces less interferences on it, increasing the angle of attack (see **Figure 15**). It is possible to find a similar trend also in scale 1:80, but the differences are minimal.

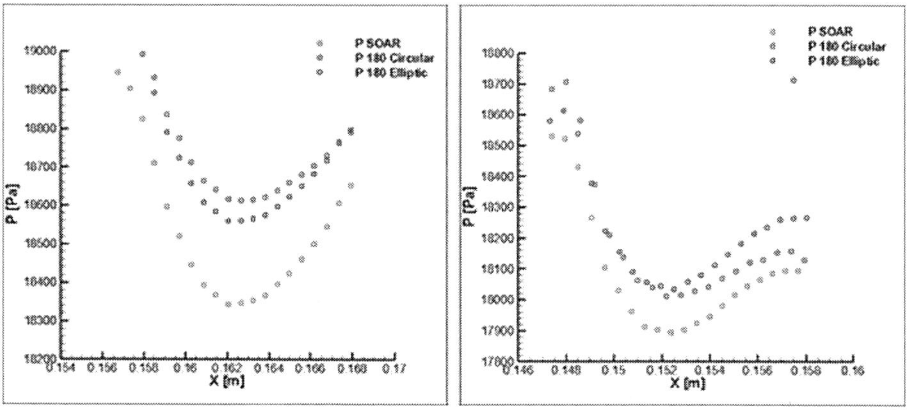

Figure 15. Pressure distribution with nominal dimensions on the flap, circular and elliptic sting; scale 1:180, α=0° (left) and α=15° (right).

4.4.2. *Coefficients corrective laws*

The coefficients coming from the simulations with support are corrected for the presence of the struts.

Scale	1:180		1:80	
Ki,j	Circular	Elliptic	Circular	Elliptic
Flap	0.989	0.990	0.996	0.997
Base	0.975	0.986	0.989	0.994

Table 4. Averaged local pressure ratio

The surface pressure alterations are considered responsible for the interferences on the coefficients, see Equation (1) and **Table 4**. The ratios ($K_{i,j}$) between the pressures, on the rear surface, of the SOAR alone divided by the same pressure when the support is present are taken into account to correct the drag. The same values on the flap surface are supposed to be determinant for the pitching moment and the lift. Afterwards from these pressure ratios a first correction coefficient is applied and then other elements are added to try to have the best fit of the simulation results.

$$K_{i,j} = \frac{P_{SoarAlone\,i,j}}{P_{SoarWS\,i,j}} \tag{1}$$

The corrected results always present a residual average error from the SOAR alone lower than 10^{-4}, and few cases have an error of 10^{-3}. The improvement from the results coming from the simulations is always at least one order of magnitude. It is useful to highlight that the nature of this correction is totally empiric.

For the drag, the pressure corrective coefficient is in the denominator because the higher pressure induced by the sting on the SOAR rear surface reduces the drag coefficient.

On the lift, the pressure coefficient multiplies the aerodynamics coefficient because the overpressure on the flap (and on the body) reduces the lift but making it less negative for α=0°.

For the pitching moment, the pressure coefficient is divided by a factor 10. This is for the huge impact that the flap overpressure has on the coefficient due to the distance from the CoG and the large surface. Then everything is multiplied to the moment coefficient because the sting effect is to increase it.

As example, just the drag and lift corrective law for the circular sting and the pitching moment corrective law for the elliptic one are reported, and all corrective laws are in Ref. [9]. All is referred to the scale 1:180.

$$C'_{D,C} = C_D \frac{1}{K_{b,C}} + 4{,}1174 \cdot 10^{-3} \quad (2)$$

$$C'_{L,C} = K_{f,C} C_L + 8.3573 \cdot 10^{-3} + 2.5311 \cdot 10^{-5} \alpha^{2,2} \quad (3)$$

$$C'_{M,E} = \frac{K_{f,E}}{10} C_M + 2.81817 \cdot 10^{-4} 8.511 \cdot 10^{-4} \alpha \quad (4)$$

5. Conclusions

All supports are structurally valid and induce an amount of interferences manageable with the suitable corrections. The conclusions about the interferences will now be analyzed more in detail.

The elliptic sting shape presents, in both scales, the best performance, especially on the pitching moment coefficient and at α=15°. For all supports with an angle of attack much higher than 10°, the usage of the elliptic sting is definitely suggested. In scale 1:180, there are evident advantages with its usage also on the drag and on the lift coefficients and also at the lower angle of attack. In scale 1:80 for all the coefficients, the interferences, in spite of the greater scale, are at least half of what happens with the 1:180 supports. For the lift and the drag, the differences between the two shapes are halved, with the same trend. The pitching moment worsening with α is reduced, especially for the elliptic sting.

Varying the dimensions, the trend is what was expected, in scale 1:180: the interferences go down increasing the length and decreasing the diameter. It is more complex in scale 1:80, as already discussed, and the range of alterations is not enough to clearly understand the behavior of the support. For the pitching moment having a sting 10% longer at 15° angle of attack dramatically reduces the interferences. The reason why this is not revealed for the circular sting in scale 1:180 is not clear yet.

Increasing the inclination of the vertical parts (only scale 1:180), the interferences increase. If the variation is 5°, the effect is minimal; at B=55.8°, the interferences rise more.

Comparing this to the old support (only scale 1:180), the first consideration is that the SOAR rear part, including their flaps, is a critical zone for the interferences. In fact, for their large flat surfaces, only small deviances on the pressure contour cause great alterations on the coefficients. This in particular affects the pitching moment for the distance from the CoG. Every kind of support will, more or less, produce alterations on the flow in this part and then a certain amount of interferences is inevitable. On the contrary, to have a support that does not interfere with the rear part is not possible. The dramatic reduction of drag and lift interferences, with a posterior sting, seems to be interesting even if it is necessary to take into account a worsening of the pitching moment. With little optimization on the design, the use of an elliptic shape and the research of some expedient to reduce the moment interferences may be the best solution.

6. Recommendations for future activities

It is interesting to examine more in depth some aspects of this study such as the impact of shape and dimensional variation. To do this, it is useful to perform simulations at more angle of attacks and to apply greater length alterations. It is possible to do this remaining in the original constraints for the scale 1:80 where they are less severe.

Furthermore, for a more in-depth analysis, it is necessary to perform CFD test of supports with the completed sustaining mechanism. This is useful especially for the straight sting (scale 1:80) where testing only the last part of the support is not possible to clarify the interference variations with the length. Therefore, it will be possible also to understand better the influence of the Reynolds number on the length.

Another critical point is to research a solution to decrease the interferences on the pitching moment. To permit this in scale 1:80, there is the possibility to position the sting higher on the SOAR rear surface. This should reduce the flow blockage over the flaps. It is more difficult to realize the same in scale 1:180 (especially for the elliptic sting) for the higher ratio dsDalready existent.

Finally, it is also interesting to estimate the accuracy and the usefulness of the new correction methods purpose in Section 4.4.

Nomenclature

α	angle of attack
B	inclination angle of the support vertical part
CoG	center of gravity
C′	aerodynamics coefficient corrected for the effect of the support
D	model base diameter
d	sting diameter
$\mathbf{K_{i,j}}$	$K_{i,j} = \frac{P_{SoarAlone\ i,j}}{P_{SoarWS\ i,j}}$
LC	critical sting length
l	sting length
M	Mach number

http://dx.doi.org/10.5772/62241

Re Reynolds number

VKI The Von Karman Institute for Fluid Dynamics

Author details

Alberto Ghiraldo[1*], Sebastien Paris[2] and Ernesto Benini[1]

*Address all correspondence to: albertoghiraldo@hotmail.it

1 University of Padua, Padua, Italy

2 The Von Karman Institute for Fluid Dynamics, Sint Genesius Rode, Belgium

References

[1] Swiss Space System. SOAR design definition file design justification file. Internal document. 2013.

[2] S Paris, BànyaiT.. Contribution to the aerodynamic database of the SOAR and the Airbus A300 during the separation—wind tunnel tests and CFD. VKI Internal report. 2014;S3-VKI-RPT-001.

[3] FB Cyran. Base Pressure Measurement of the Standard Dynamics Model Aircraft at Mach Numbers 0.3 Through 1.3. Tennessee, USA: Arnold Air Force Station; 1981.

[4] A Pope. Wind Tunnel Calibration Techniques. AGARDograph; 1961.

[5] S Hoerner. Fluid Dynamic Drag; 1965.

[6] HM Schurmeier. An Investigation of the Interference Effect of a Sting Support System on the Pressure Distribution over a Body of Revolution. Pasadena, CA, USA: Pasadena Institute of Technology; 1949.

[7] JD Anderson Jr.. Fundamentals of Aerodynamics. 5th ed. New York: McGraw-Hill; 2011.

[8] DL Loving, AA Luoma. Sting Support Interference on Longitudinal Aerodynamic Characteristics of Cargo-Type Airplane Models at Mach 0.70 to 0.84. Langley Station, Hampton, VA, USA: Langley Research Center; 1967.

[9] A Ghiraldo Effetto del supporto in misurazioni in galleria del vento (in English; only abstract available). Padua thesis. 2015.